Commercial Solar

Step-by-Step

COMMERCIAL SOLAR

Step-by-Step

A Series of Dialogues and Discussions

JIM JENAL, FOUNDER & CEO, RUN ON SUN

Cover photo by Andrea Bricco and used with permission.

Cover design by Julia Jenal.

For more information, contact info@runonsun.com, or visit

www.RunOnSun.com.

ISBN-13: 978-0615843766 (Run on Sun Publishing)

ISBN-10: 061584376X

First Edition: September 2013

DEDICATION

To Julia, my Purpose.

Contents

LIST OF FIGURES

FOREWORD

Solar energy is here to stay. However, investing in a solar PV system for a commercial building can feel like a big step, and the breadth of considerations can be overwhelming. *Commercial Solar: Step by Step* lays out the process of research, analysis, and implementation needed to realize project goals. The material could be dry (much of the reading on this subject is), but is instead casual but precise, clearly laid out, and made accessible through handy use of a narrative in which the Facilities Manager of a fictional company undertakes a commercial solar project himself.

Commercial solar power systems are expected to be the fastest-growing segment of the PV industry in the coming years. Understanding the impact of a PV system on a balance sheet and P&L, accounting for cash flows, tax benefits, rebates, and other subsidies, choosing the right financing and partnering with the right contractor are all essential to defining and implementing a scope of work that achieves the desired outcomes. As more businesses ascend this learning curve to internalize the benefits of solar, the pressure is on their competitors to do the same. Executives and other stakeholders in business need to get comfortable with solar power projects and their implications, and this book is a valuable contribution towards that end.

In the three years that I have known and worked with Jim Jenal, he has become highly respected in the solar PV industry for his insightful commentary on industry issues, which can be seen on his blog, on his website, and now in this new book. In his typical style—approachable, honest, quirky, and occasionally scathing—Jim has thoughtfully flattened out the complex world of commercial solar PV into an understandable roadmap that anyone can follow to project success.

Boaz Soifer
Vice President of Sales
Focused Energy LLC

PREFACE

Solar power is booming across the U.S.—solar installations accounted for 48% of all the new energy production capacity installed during the first quarter of 2013[i]—but for many people it remains a mysterious and even daunting subject. Sadly, this is especially true for those who could benefit the most from installing solar: the owners and operators of commercial buildings.

Faced with ever higher costs from electric rates that are constantly increasing, producing your own energy to save money certainly *seems* attractive. But how can a harried facilities manager find reliable answers to the myriad questions that present themselves: How can you identify qualified contractors? How do you go about assessing their competing bids? How should your company pay for it? And how can you even get started when you don't know what questions to ask in the first place?

If these questions sound familiar, this book is for you.

In the pages that follow you will meet Jack Prince, facilities manager for expanding bio-tech company EnGex, who suddenly finds himself tasked with determining whether his company should "go solar." Through a series of dialogues between Jack and his colleagues that frame the issues, followed by subsequent discussions that provide greater detail, you will follow Jack as he learns all that he needs to know about commercial solar power systems: from mastering the basics to commissioning the installed system. And, because in the real world the story doesn't end there, we will also present two case studies about actual commercial solar installations a year after the fact, but with very different endings.

My hope is that this book will help demystify the process and clarify the benefits of going solar, step-by-step.

After years as a scientist, educator and trial lawyer, I founded Run on Sun in 2006 because I believed that there was a niche to be filled by a solar company run by people who could answer a client's questions in a clear and concise manner. I knew from my years as a scientist at Bell Labs that technical details mattered, but I also knew from teaching that the "curse of knowledge" left many "experts" incapable of explaining technical subjects to an educated, albeit lay, audience. And I knew as a trial lawyer that sophisticated decision makers—whether Judges or CEOs—had well developed abilities to sense "the ring of truth," even if they didn't understand the subject matter well enough to know what the "truth" was.

In the years since I have come to believe that all of those considerations are even more important today than they were seven years ago.

This book is the result of that experience. I hope it helps.

Jim Jenal

Pasadena, California, September 2013

Chapter Notes

[i] *Solar accounts for 48% of new electric capacity in Q1 2013,* from U.S. Solar Market Insight Q1 2013 by GTM Research/SEIA®, available at http://www.seia.org/research-resources/us-solar-market-insight-q1-2013, accessed 6/16/2013.

I. INTRODUCTION: WHY GO SOLAR AT ALL?

MEET JACK PRINCE

Jack Prince looked at the phone on his desk and scowled. It wasn't that it was ringing—Jack's phone was always ringing. No, it was the name that popped up on his caller-id screen that was causing his consternation. Amy Peller, CFO at EnGex, was on the line, and Jack was certain this wasn't a social call.

"Hi Amy," Jack answered. "What's up?"

"Have you seen these bills?" Amy demanded in her typical, no-preamble approach to conversations. Amy always knew exactly what was on her mind. Unfortunately, her subordinates only occasionally did. This was one of those awkward moments for Jack.

"Which bills do you mean?" he asked, knowing even as he did, that it was the wrong question to ask.

"These electric bills, Jack!" Amy exclaimed. "They are completely out of control. What are you going to do about them?"

Jack sighed. As facilities manager it was his job to be on top of operating expenses for the five buildings that comprised the EnGex campus, and he had made substantial efforts to limit their energy usage even as the company grew. But recent rate increases for the local power company—seventeen percent over just the last three years—threatened to overwhelm his best efforts to date. From the edge in her voice it sounded as if Amy had reached the same conclusion.

Jack considered his words carefully.

"I think we have already picked all of the low-hanging fruit. If we are really going to make a dent in our electric bills, we are going to have to take a more aggressive approach."

Now it was Amy's turn to sigh.

"Look, Jack, I know you are good at what you do and heaven knows we would be way worse off if it weren't for all the clever changes you've introduced over these past five years. So please, don't get me wrong—I'm not faulting you. It's just that we need to have a plan that will give us some assurance that we can get these costs under control.

"I need you to figure this out—and soon. Our annual meeting is in six weeks and Jason is going to want us to give him something concrete to bring to the Board."

Six weeks. Jack's stomach turned—not a lot of time to come up with something "concrete" to solve the problem of higher electric bills.

"I'm on it," he said.

"Good, for both our sakes," replied Amy, hanging up.

Jack knew that Amy was right. In the nearly six years that he had been facilities manager at EnGex he had only spoken one-on-one with CEO Jason Loudon a handful of times but the impression was indelible: Loudon was a bottom-line kinda guy. If something didn't pencil out economically, it was a non-starter with him. That said, he willingly supported good ideas and rewarded success handsomely. Failure, on the other hand, received an equally vigorous—if markedly less pleasant—response.

Jack would have to get creative; the easy things had already been done, so now what? Jack's stomach twisted a little more.

As he sat there trying to formulate a plan, one presented itself to him by way of yet another phone call, this time from Janet Lang, Director of Engineering.

"Hey Jack, I know you are busy, but I just saw something pretty impressive, and I thought of you. Do you have a second?"

"For you, sure," said Jack. Janet was an ally and very smart. Her insights had proved helpful in the past. "What's on your mind?"

"I just came from the company where my daughter Lena works, and she was really excited about what they are doing there. Apparently they just installed a solar power system on their warehouse roof, and she told me that they expect the system to pay for itself in four to five years while saving them something like sixty-five percent on their energy bills every year.

"Anyway, I know you are always looking for ways to improve things around here, so I thought I would mention it to you."

"Oh, I don't know," said Jack. "We looked at solar a few years ago and it just didn't pencil out."

"How long ago was that, Jack?" asked Janet. "According to Lena, they were really skeptical at first, but the economics of it really wowed them."

"Well, now that you mention it, it was probably close to ten years ago. Do you really think that could work for us?"

"I don't know, but ten years is an awfully long time when you are talking about a high tech industry like solar. I can find out from Lena who they used at her company. What have you got to lose?"

"Nothing," Jack said, half to himself. "Nothing at all. Thanks, Janet. As always you've made my day. Please ask Lena for the contact information, and thanks again."

For many facility managers and building owners alike, Jack's problem sounds all too familiar. Jack's boss, like many a boss trying to run a successful business, is focused on the bottom line and rising operating costs are a cancer, eating away at the company's profitability. They are well aware of rising energy costs and they have already done "the easy things"—upgraded their lighting, replaced antiquated air conditioning equipment with more efficient HVAC systems, even installed "smart controls" in an effort to reduce their peak power demand—but still their monthly bills rise. They need a new approach, and they need it now.

It is the central premise of this book that a commercial solar power system is just what Jack and his boss are looking for to get out of their present predicament. And to be sure, while there are numerous reasons for adding solar beyond the economics, few companies can commit to going forward if the economics don't make sense.

In the coming pages you will follow Jack as he progresses, step-by-step, to learn what he needs to know to understand how solar can help EnGex save real money. He will learn what is hiding in those ever rising electric bills, and he will figure out how to locate not just one, but several reliable solar contractors to provide him with comprehensive proposals.

We will watch as Jack sifts through the competing proposals and see how he figures out how EnGex should finance their system. Then, with a contract signed, we will learn from Jack's experience in overseeing the installation process from pulling permits to processing rebates.

Jack doesn't know it yet, but he is about to become an expert on how to get a commercial solar power system installed.

And so are you.

II. PRELIMINARIES: WHAT YOU NEED TO KNOW FIRST

JACK DOES SOME HOMEWORK

Like water, trouble also flows downhill, and Jack's next call was to Terri Flint, a local energy efficiency consultant who had been his go-to person on the HVAC replacement project EnGex did two years ago. Jack figured that her expertise could come in handy now.

"Hi Terri, it's Jack, and I really need your help."

"Sure, Jack, what can I do for you?" Terri asked.

Jack recounted his conversations so far. "I was hoping you could give me some ideas on how to get started," he concluded.

"Absolutely," Terri replied, "happy to help. Let me ask you this—how much do you actually know about your electric bills?"

"Well, I know that they are too high to please my CFO!"

"Right—no surprise there. But have you actually looked at how your bill is computed?"

"Not really," Jack confessed, feeling a bit sheepish. "I looked at them a few years ago after I read an article about commercial customers being routinely overcharged. But I really had a hard time figuring out how they came up with the numbers that they did. Does anyone actually understand those bills?"

"Sure," Terri laughed, "we do. And a reputable solar installation company certainly should. In fact, if you talk with a solar company and they can't explain exactly what is going on with your current bills, you need to talk to a different solar company!"

"Good point," said Jack, "I'll keep that in mind. But right now I don't even know what questions to ask—I guess I don't know what I don't know."

"Ok, let's go over some basics," Terri agreed. "Most commercial utility customers are charged for electricity in two ways: usage and peak demand. Usage is easy; that is how you are billed at home—the more energy you use in a month, the bigger your bill gets.

"But demand is sneaky. It is based on the highest spike of power that your building draws during the entire billing cycle, and it can really increase your overall bill."

"Are you saying that turning everything on at the same time every morning might actually be increasing my electric bills?" asked Jack.

"Without a doubt," Terri replied. "Your solar installer should look over a year's worth of your bills and be able to explain to you where your money is being spent. They should also have detailed models that actually track the rates charged by the utility so that they can give you an accurate estimate of how much your solar system will save you."

"Okay, what else do I need to know?"

"Are you familiar with net metering?"

"I've heard the term before," said Jack. "Doesn't it have something to do with how you get billed when you have a solar power system?"

"That's right. Under a net-metering agreement, the utility will install a new meter that can measure and record energy flow in both directions. When the solar power system produces more energy than you are consuming with your loads, that excess energy will flow out onto the grid and you will get a credit for it. At night, or on a stormy day when the system isn't producing much energy, or none at all, you will draw energy from the grid to power your loads, and the meter will keep track of that as well.

"At the end of each month, the utility nets those two numbers out. If you consumed more than you produced, you are a net energy *importer* and you will pay for that balance. On the other hand, if your system is large enough so that you are a net energy *exporter* at the end of the billing cycle, the excess will be carried forward as a credit to the next cycle."

"Ok, what else?"

"You will want to know about rebates," said Terri. "Most utilities will pay rebates for commercial systems based on the actual amount of energy that your system produces. Those rebates are generally paid over a number of years, most commonly five."

"Oh, but Lena's neighbor just put solar on his house and he got his rebate all at once," protested Jack. "I know because he made a big deal out of bringing over the check and bragging about it to her."

"For residential and really small commercial systems, rebates are paid out in one lump sum when the system is commissioned," Terri explained. "But for larger systems, like what you will need for EnGex, the utilities like to spread out the payments over time – and only pay for the power that actually gets produced."

Jack nodded. "I guess I can't fault that logic. I wouldn't want to pay for something I didn't get either. Anything else I need to keep in mind?"

"The last thing is a bit out of my realm, but you should talk to the company's tax accountant about the available federal and state tax incentives. The most significant one is the thirty percent federal Investment Tax Credit, but there are also federal and state depreciation formulas that may apply. As I said, that's a bit out of my depth, but your accountant can explain that to you."

"Wow, thirty percent? I had no idea. I'll walk over to accounting and see if someone can give me the details. This has been really helpful. Thanks, Terri."

"No problem at all, Jack. Good luck."

As many companies sit on the sidelines with accumulated capital, spending some of that capital on a commercial solar power system often makes great economic sense. But for some companies (and their facility managers, accountants, Board Members, *etc.*), commercial solar is still a confusing concept, filled with impenetrable jargon and competing claims. Can a commercial solar power system really be as economically beneficial as the proponents claim?

Rather than answer that question directly, it is useful first to lay out the case for commercial solar power in some detail. Although no chapter in a book can take the place of a face-to-face conversation that takes into consideration all of the unique elements of a specific company's situation, there are enough common elements that can and should be explained to demystify the overall process before that conversation ever takes place.

First things first. Before you ever call a solar power company—and we will explain how to find the good ones later in this chapter—you need to start with something more mundane: your electric bill. When was the last time that you really looked at your electric bill? For many business or building owners the answer is never. Oh sure, you certainly know how much you are paying—but do you know *why* you are paying so much? What horrors are hiding in your electric bills?

UNDERSTANDING YOUR ELECTRIC BILL

The Basics—Usage & Demand

There is probably a very good reason why neither you nor anyone else at your company has ever looked closely at your electric bill—it is terribly confusing. Let's start with some basics. Almost every commercial user pays for two major components on their electric bill: **usage** and **demand**.

Usage is the more familiar component as it is the basis for your residential electric bill. It is based on the total amount of energy that you used over the course of the billing cycle (usually one month for commercial customers). Usage is measured in total kilowatt hours (kWh). Usage charges are based on some specific cost per kWh as defined in the rate schedule that applies to your utility account (more on rate schedules in a moment). Depending on the design of that rate structure, it could either be a fixed amount (that is, every kilowatt hour costs the same), a variable amount tied to when the energy was consumed (either time of day or time of year or both), or a variable amount based on levels, or tiers, of total usage.

Demand is a bit more complicated—it is usually defined as the greatest amount of power that the utility has to provide to you over a measured period of time during the billing cycle. For customers of California's Southern California Edison (SCE), demand is the peak power required during any 15 minute period over the month. That means that if your building has multiple HVAC units and they all come online during the same 15-minute window, your demand will spike much higher than it would if those units came on in a staggered fashion (since the power demand of an HVAC unit is highest when the unit is first started.) Demand charges are billed per kilowatt (kW) of power and, for many commercial customers, ***demand charges may account for more than half of your total bill!*** As with usage, how you are charged for peak demand depends on the design of your applicable rate structure. Demand charges might be a fixed amount per kilowatt needed, or it may be tied to the time of use, or even tiered.

Rate Schedules

Rate schedules, or *tariffs,* as they are known to electric utility regulators, are the formulas that a utility uses to calculate your bill. Every utility has a multitude of rate schedules that might apply to a commercial building[1] and you could pay vastly different amounts—that is to say you could save a lot of money—by switching to the most economical rate schedule for which you qualify.

Case in point: SCE has two rate structures that commonly apply to small to medium size commercial buildings: **GS-1** and **GS-2**. The beauty of the **GS-1** rate schedule is that it has no demand component. But here's the catch—your peak demand must not exceed 20 kW in any three of the previous twelve months. Once you exceed the 20 kW demand threshold you are in the realm of **GS-2** which adds a substantial demand component. In fact, it adds two: one that applies only during the summer months (actually from June 1 to September 30) and a second that applies year round. That means that during the summer, you are paying for every kilowatt of demand—twice. Ouch.

We had one potential client whose bills revealed that they were paying under **GS-2**. When we analyzed their bills—the first step in preparing a proposal for installing a commercial solar power system—it was apparent to us that they were actually entitled to be billed under **GS-1**. When we met with their facilities manager to discuss our proposal, we pointed out that they ***could have saved over $2,000 the past year*** if they had been on the right rate schedule. We encouraged them to contact SCE about getting switched to **GS-1**. (No, SCE had not suggested that to them.) Strangely, none of the other solar companies that they had talked to had explained that to them, yet once they called SCE, they were switched over immediately.

Here's another example. One of the local municipal utilities in Southern California, Pasadena Water and Power (PWP), generally has low rates, but their mid-level commercial rate schedule (**M-1**) has one of the most significant "gotchas" we have seen anywhere—and we are yet to speak to a single potential client who was aware of this before we pointed it out. The **M-1** rate structure includes a demand component (denoted as a "distribution" charge), but unlike SCE's demand component described above, PWP charges you for the ***peak demand in any 15-minute window for the past 12 months!*** That means that if on one unlucky day, everything in your building comes online all at once during the same 15 minutes, not only will you pay for that peak demand that month, you will pay for that peak demand for *every* month for the next year (unless a higher demand comes along to take its place). For one of our clients, they had a peak demand one month that spiked at 82 kW, yet their average for the next 12 months was only 36 kW. Under the M-1 rate schedule, they paid $5,300 more than they would have if they only paid for their monthly peak demand.

Gotcha, indeed!

Models Matter

One of the most important tools a solar company can offer its potential client is a properly designed and *up-to-date* set of rate schedule models that allows them to analyze accurately a company's prior utility bills. From that analysis it is then possible to make reasonable predictions regarding potential savings based on a host of measures: changing rate schedules, reducing usage or peak demand, ***and/or*** installing a solar power system.

Unfortunately, some solar companies simply assign a fixed amount of savings per kWh to their proposed solar power system's annual energy yield and call that your potential savings. Such an approach ignores the complexities of how your electric bill is actually calculated and may overstate the financial benefit of adding solar.

So before you pick up the phone, pick up your electric bills and check out what is hiding there—it is the first step in getting the greatest value from your commercial solar power system.

UNDERSTANDING NET METERING AND REBATES

Net Metering

In most grid-tied solar power systems, power from the utility's grid is provided at night and on cloudy days when electricity demand in the business exceeds the power generated by the system.

During the daytime when the power generated by the solar power system exceeds the local energy demand, the excess electricity is delivered to the utility grid thus "spinning the meter backwards." This is the concept of "net metering," whereby the excess energy you generate throughout the year offsets the energy you consume.

California law requires utilities (except LADWP) to pay owners of solar power systems for any excess energy that they produce. While most solar power system owners will not be net energy producers, this change in the law is an important step in making net metering agreements more equitable for system owners.

There is an alternative to net-metering, known as a Feed-in Tariff, that may be applicable to some commercial solar installations. Because Feed-in Tariffs are still rare in the United States (although they predominate in solar-friendly countries like Germany), we will postpone our discussion of them until Chapter VI. Special Cases.

Rebates

In most utility regions, owners of commercial solar power systems are paid a rebate directly from the utility. A solar rebate is an assignable payment to the utility customer as an incentive for installing the system. These rebates are typically funded by a special fee assessed on electric bills such as a "public benefit" charge.[2]

Regardless of the source of funding, there are two generic types of rebates used by utilities: Expected Performance Based Buydown (EPBB) and Performance Based Incentive (PBI).[3] While different utilities may offer one or both types of rebates, for any given system only one type of rebate will be allowed.[4]

EPBB Rebates

The EPBB rebate is a one-time, lump-sum payment based on the ***expected*** (as opposed to the measured) production of the solar power system. An EPBB incentive is generally limited to residential and small commercial/non-profit/government installations of 30 kW or less.

These rebates are calculated based on the anticipated output of the system. In California they are typically based on what are known as CSI (for California Solar Initiative) AC Watts. To determine that value, the CSI rebate calculator is used, and its calculation takes into consideration the performance characteristics of the equipment selected (*i.e.*, PTC[5] rating of the solar modules and the efficiency rating of the inverter(s)), the azimuth and pitch of the array(s) (which determine the array's orientation toward the sun), the geographic location of the system site, and any shading factors that might affect the system's anticipated performance.[6]

EPBB rebates are a good deal for system owners since they provide all of the rebate money upon commissioning of the system. However, they are a decidedly less beneficial arrangement for the utility, particularly for larger systems. Since EPBB rebates are not tied to actual performance, the utility pays the same rebate for under-performing as well as over-performing systems.

PBI Rebates

PBI rebates are intended to overcome those problems and insure that the utility is not providing an incentive for systems that fail to meet expectations, or don't operate at all. A PBI rebate typically consists of five annual payments based on the actual performance of the solar power system. PBI rebates are calculated based on a fixed price for energy produced, expressed in cents per kilowatt hour, in a similar way to how usage is calculated on your electric bill. As a result, the more energy the system produces, the higher the rebate. Unfortunately, if the system does not live up to its projected annual production, the system owner will have the unpleasant surprise of receiving a lower than expected rebate.

In some cases, utilities will allow a customer to opt into a PBI rebate if their system size is in a certain range, frequently between 10 and 30 kW, but generally all systems above 30 kW will be required to receive a rebate under the PBI method.

One important difference with a PBI rebate is the need for an approved performance monitoring system so that the utility (along with the customer) knows exactly how much energy the solar power system is producing each month, since that is the basis for the rebate payment. The complexity of that monitoring system can vary widely, from simply installing a revenue-grade meter (that is, a meter that measures energy production to the utility's specified tolerance, such as ± 0.2%), to a complete remote monitoring and reporting system supplied by a third-party vendor.

Since the utility only pays for power actually delivered, rebate dollars are guaranteed to provide the bargained for benefit. However, because of the potential need to supply the utility with verified performance data, PBI rebates may increase the Operations & Maintenance (O&M) expense of a commercial solar power system, at least for the five years of the PBI rebate. On the other hand, if your system is well maintained and conservatively designed, you may actually receive more in rebate payments than originally projected.[7]

UNDERSTANDING TAX INCENTIVES

Federal and state tax incentives are another important concept to understand in the economics of a commercial solar power system. Although a detailed discussion of the tax incentives available to a particular project is a topic beyond the scope of this book, there are two broad classes of tax incentives that will apply to most commercial solar power systems that are purchased by for-profit entities: the federal Investment Tax Credit and Accelerated Depreciation.

Federal Investment Tax Credit

The federal Investment Tax Credit (or "ITC") provides a full 30% of the direct cost of the solar power system in the form of a tax credit to the entity that owns the system. The applicability of the ITC to indirect costs—such as deciding to re-roof your building before adding solar—must be decided on a case-by-case basis.

Since commercial rebate payments are generally treated as taxable income for a commercial building owner, the value of the rebate is not deducted from the system cost when calculating the basis for the rebate. (This is the opposite of what typically occurs for a residential solar client since residential rebate payments are generally not taxed; hence, the value of the rebate must be deducted from the system price in calculating the basis for the ITC.)

Since the ITC is a one-for-one reduction in the amount of tax actually owed, it is possible that it might exceed the system owner's tax liability in the year that it is earned. Fortunately, IRS rules allow the excess credit to be carried backward one year (by filing an amended return for the previous year) and forward up to twenty years.

The ITC is scheduled to step down from 30% to just 10% starting in 2017.

Accelerated Depreciation

Another important tax incentive is depreciation, by which a taxpayer can deduct the cost of purchased equipment against their net income. As this is written (Summer 2013), federal law allows an accelerated depreciation schedule to be applied to solar equipment, and some states—such as California—do as well.

Since depreciation reduces the taxpayer's adjusted gross income, its cash value is dependent on the taxpayer's overall tax bracket. However, for high-income taxpayers, depreciation can significantly reduce the overall cost burden of adding solar.

As these can be very complicated questions to answer, please consult with your tax advisor for the specifics that might apply to your situation.

Chapter Notes

[1] See, for example, the tariff page of the SCE website which lists some twenty-two Commercial rate structures, each of which contains multiple sub-types. https://www.sce.com/wps/portal/home/regulatory/tariff-books/rates-pricing-choices/business-rates, accessed 6/2/2013.

[2] In California, public benefit charges are used to pay for a variety of utility programs, including rate reductions for customers on fixed-incomes.

[3] A database of rebates for solar power systems nationwide, compiled by North Carolina State University, is available at the DSIRE website, http://www.dsireusa.org/, accessed 6/2/2013.

[4] Some utilities will allow the system owner to elect which type of rebate they will receive.

[5] PTC stands for **P**hotovoltaic **T**est **C**onditions and is meant to indicate the "real world" power output of the tested solar module. As a result, it is always lower than the nameplate rating of the module.

[6] For more information about how the various equipment and environmental factors cited here affect the performance of the system, see Chapter III. Moving Forward: Comparing Contractors & Their Bids - Apples to Apples, p. 29.

[7] As was the case for the system installed at Westridge School for Girls. See Chapter VII. Lessons Learned: Reporting from the Rooftops - Second Case Study: Westridge School for Girls, Pasadena, California – The Beauty of Doing it Right, p. 117.

III. MOVING FORWARD: COMPARING CONTRACTORS & THEIR BIDS

JACK'S QUEST FOR SUITABLE SOLAR CONTRACTORS

Jack was on the phone again.

"Hey Terri, it's Jack. I've done my homework, I think, and I now have a year's worth of bills scanned and ready to go. But I've got a big problem: there must be a thousand solar companies out there. How should I go about picking one for us?"

"That's a great question, Jack," Terri agreed. "If you haven't ever done this before, how can you know who you can trust to do it right?"

"Exactly," said Jack. "If we are going to do this, it has to be a complete success. I don't need to go with the lowest bid, as long as I am getting real value for the money. But I really have no idea how to start. I've heard some companies advertising on the radio, or online, and I've gotten a bunch of junk mail at home—heck, I've even had people coming door-to-door like they were selling vacuum cleaners or aluminum siding! It's kinda crazy."

"Doesn't inspire a lot of confidence, does it?"

"No, quite frankly, much the opposite. Still, I'm sure there must be reputable companies out there—so how do we find them?"

"Well, Jack, I always say you want to find a solar contractor who is NEARLY perfect."

"What do you mean, nearly perfect? I don't even know where to start, let alone how to find someone who is perfect – though I agree that would be nice."

Terri laughed. "Not *nearly* perfect, Jack, NEARLY. It's an acronym. It stands for:

N – NABCEP
E – Electrician on staff
A – Experience with the AHJs
R – References for similar-sized projects
L – Local or national?
Y – Years in business

"**N**ABCEP is the leading accreditation organization in the solar industry, and you really have to know your stuff to get certified. They have a list of certified installers on their website—so that is a really good place to start.

"**E** is for Electrician, and you want to make sure your prospective solar contractors have a properly licensed electrician on staff.

"You should also check with your local **AHJ,** the **A**uthorities **H**aving **J**urisdiction over your project, like the city Building department and the Utility. They often maintain a list of approved contractors. But even if they won't recommend anyone, they might warn you of people who they think are less than reliable.

"**R**eferences are particularly important, and even though you are only likely to be given ones that will say great things, you want to be sure to contact the references for projects of similar sizes.

"The last two are a little more a matter of personal taste and comfort: is it a **L**ocal or national company and how **Y**ears have they been installing commercial solar systems? Personally, I like to work with local companies when I can, and I wouldn't do a project of your size with a company that hasn't been in business for at least five years. That tells me they can manage their finances as well as design solar power systems.

"Put it altogether, and you have a contractor who is NEARLY perfect."

"That's great," said Jack. "I love things that are easy to remember, and NEARLY perfect fits the bill. Thanks again."

"You're welcome, Jack. Happy hunting!"

FINDING YOUR "NEARLY PERFECT" SOLAR CONTRACTOR

Before you can ever get a bid, you have to contact a solar installation contractor to come out to your location and perform a site evaluation. Actually, you should contact at least three contractors so that you have a set of bids to compare—but how do you find them in the first place? Well, you could choose based on who has the most ads on TV or the Internet, or you could rely on Cousin Billy's recommendation, but somehow that just doesn't seem sufficiently *scientific* for a project like this. There has to be a better way, and there is.

As Terri explained to Jack, you want to find a contractor who is NEARLY perfect. Fortunately, that isn't as hard as it sounds.

NABCEP Certification

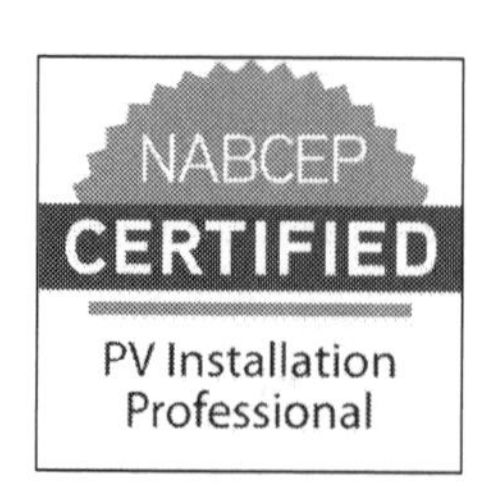

Figure 1 - NABCEP Certification Seal

The North American Board of Certified Energy Practitioners[1]—NABCEP for short—provides the most rigorous certification process of solar installation professionals in the industry. Not to be confused with their Entry Level Letter that merely demonstrates that the person has taken an introductory course in solar, the *NABCEP*

Certified PV Installation Professional credential is the Gold Standard for installers and consumers alike. Earning NABCEP Certification requires the successful candidate to have an educational background in electrical engineering or related technical areas (such as a Union apprenticeship program, for example, like what is offered through the IBEW[2], or military training), at least two solar installations as the lead installer, and the successful passing of a 4-hour written examination on all aspects of solar power system design and installation.

As NABCEP explains:

> When you hire a contractor with NABCEP Certified Installers leading the crew, you can be confident that you are getting the job done by solar professionals who have the "know-how" that you need. They are part of a select group of people who have distinguished themselves by being awarded NABCEP Certified Installer credentials.[3]

NABCEP's website offers a database of all Certified Solar PV Installers—just enter your zip code to find the installers located near you. It is with great pride that we point out that at Run on Sun, all three of our owners have achieved the designation of NABCEP Certified PV Installation Professional, and we know of no other solar power company in Southern California that can make that claim.

Licensed Electrician on Staff

Commercial solar power systems are complicated electrical systems that involve high voltages and currents. To complete such a project safely requires the involvement of a licensed electrician (a C-10 license in California), preferably one who has experience working with commercial/industrial electrical systems.

Unfortunately, not all solar contractors have licensed electricians on staff. In California there is a so-called "solar contractor" license (C-46), which was originally created for installing solar hot-water systems back in the 1970's. All fine and good if you want to heat your swimming pool, but a contractor with such a license can also install PV systems with their high voltages and large amounts of power. Ask yourself this: would you hire a plumber to rewire your home? Probably not. All the more reason why hiring a C-46 for a commercial solar power project is a bad idea.

AHJ – Experience with the Authority Having Jurisdiction

Solar projects require the contractor to work with at least two AHJs—the local Building department (possibly also including the local Fire department) and the local utility. You want to find a contractor who has experience with both. That will help insure that your project goes smoothly without unanticipated delays or costly corrections.

AHJs can also be a source for information about a potential installer. For example, the *Go Solar California* website for the California Solar Initiative has a database of CSI-listed solar installers.[4] Every installer who has done a solar power project for a CSI utility (*i.e.*, SCE, PG&E or SDG&E) will be included on this list. Unfortunately, there are no other criteria associated with getting listed, and there is really no verification done by CSI to guarantee that the listed installer is reliable. If your job is in California, your contractor must be on this list. However, this is a double-check only—not a starting point for your search.

Another source for information about solar installers is your local utility's point person for solar rebates. This person deals with installers on a daily basis, and while she won't give you a specific recommendation, she may be able to warn you off of an installer she has learned is less than reliable.

Similarly, the professionals in your local building department deal with installers regularly as part of the permitting and inspection process. Once again, they most likely won't be in a position to provide referrals, but they may be able to give you a warning if there are red flags associated with a contractor that you are considering.

References

You would not make a critical hire at your company without contacting the prospective employee's references. You would want to know everything that you could about this potential new addition to your team. Did they execute on the assignments given? Were they proactive in finding solutions to problems? Were they honest? How did they handle mistakes when they occurred? Were they a team player?

Most significantly, you would want to apply this information to a work setting that is as close to yours as possible. After all, you will be investing a great deal of time and money in this new hire and you want to make sure that you are getting your money's worth for the long haul—you want them to be a "good fit."

All of those concerns are amplified as you investigate potential solar contractors. Your upfront investment is much greater, and you are trying to create an asset with a twenty to twenty-five year useful life. Long haul, indeed.

Glowing reviews from past clients are essential. After all, the contractor selected the references that you are contacting, so they *should be* only saying great things. But as the mathematicians would say, glowing reviews are a *necessary*, but not *sufficient*, condition to satisfy your inquiry.

Rather, you want to press the references to determine how similar their project was to yours. It doesn't tell you much if you are hiring someone to do a commercial project and the only references you receive are from residential projects. (Of course, everyone has to start somewhere and you might be comfortable being this contractor's first commercial client—but at the very least it raises a warning flag.)

Similarly, if you are looking to have a 1.5 Megawatt solar farm installed in the desert, you might not be satisfied with a contractor whose largest project to date was 100 kW on a commercial building's flat roof.

Receiving references closely related to your project's anticipated size and scope is your best indicator that the potential contractor has the experience to handle your project with as few surprises as possible. Skip this step at your peril.

Local or National?

Solar installation companies come in all sizes—from national organizations that have crews installing systems all across the country, to local operations that only work in a limited geographic region. To be sure, there are pluses and minuses associated with either end of the size scale. For example, you might receive lower prices from the national chain due to economy of scale in their purchasing, but experience delays in getting the system installed.[5] Worse still, the crew that performs the work might have been assembled from first-time, low-skilled laborers. Contrast that with the greater attention to detail you can get from a local company that lives or dies based on how well it satisfies its local customer base, albeit at a possibly higher upfront cost.[6]

Of course, money spent with a local company tends to stay in the local economy—another consideration in tough economic times.

Years in Business

The last of the NEARLY perfect elements is looking at the number of years the company has been in business. Again, this is not a perfect indicator—some recently incorporated ventures really have their act together, whereas some long-standing companies have long since ceased to care about what they are doing—but at a minimum you want some assurance that the people you are doing business with know how to run a business. Otherwise you run the risk of having no one to call if things go wrong and being stuck with a largely useless "ten-year" warranty.

We would recommend a minimum of three-to-five years in the business of doing solar, with preferably a longer track record of running one or more businesses. Expertise in areas beyond just installing solar is also useful such as engineering, management and law.

MAKING THE SITE EVALUATION WORK FOR YOU

Congratulations!

You have identified three companies to bid on your commercial solar project—what should you expect from the process? Here are three necessary steps that an installer ***must do*** to provide you with a proper proposal. If any of these steps are missing, be very skeptical of the resulting bid.

1. ***They will review a year's worth of your electrical bills***—Every proposal begins with an assessment of your needs, and that can only be determined by reviewing your electric bills. While some utilities allow you to download your usage and costs from their websites, that online information is rarely as detailed as the full bill itself. So plan on having copies ready—or better yet, scan them, have them in a PDF file, and be prepared to email them to your prospective solar contractors when you make your initial contacts.

2. ***They will go on your roof***—Nothing takes the place of actually walking your roof, taking proper measurements, and identifying any shading issues that you might have. While many commercial roofs appear to be "unshaded", a solar professional knows that appearances can be deceiving and will not rely on a client's claims but will instead do a proper shading analysis. If you have any questions about this, demand that the installer provide you with the output from their shading analysis. If they didn't perform a

shading analysis, that is another red flag. Remember—your utility requires a shading analysis as part of the rebate process, so if your potential installer didn't do one, how can they know what your rebate will be?

(Of course, there are other options for where to locate your solar array, and we will discuss one of them in Chapter VI. Special Cases.)

3. ***They will check out your electrical system—*** One of the major variables in commercial solar is the nature of the interconnection between your solar power system and your existing electrical infrastructure. Although in a residential setting this is usually nothing more than adding a circuit breaker to your existing service panel, in a commercial setting this may involve transformers, 3-phase systems and/or line-side taps, all of which increase the cost of the installed system. A professional installer will review the interconnection issues when they visit your site and adjust their bid accordingly.

Demanding a proposal that incorporates all three of these components is your best guarantee that the proposal will reflect the actual cost of the project and help you to avoid inconvenience, delays and costly change orders once the project is underway.

ASSERTING CONTROL OVER THE BID PROCESS

Apples to Apples

Getting multiple bids is essential, but understanding those bids so that you can make a true, apples-to-apples comparison can be daunting. But don't forget, you are in charge here and you should demand that your potential solar contractors take the time to answer your questions—*all of them*—so that you can make an informed decision.

One good way to start is to have the installer come in and present their proposal to you (and any other decision makers on your team). A professional installer should be happy to spend some quality time with you and your team to explain the proposal that was given to you, but keep in mind that they too are busy people. Do your homework first: compile your questions and those of your team (if they will not be participating in the meeting) so that your time together can be as productive as possible.

As part of your due diligence, here are some things to look for as you start to compare your bids.

Solar Modules

Solar modules, also referred to as solar panels, are not all alike, and although they may seem like a commodity to you, there are a number of ways in which one "200 Watt Solar Module" will differ from another. Here are the key considerations:

1. ***Efficiency***—The efficiency of a solar module tells you how much nameplate power per square foot the module will deliver. As a general proposition, at solar noon (when the sun is at its highest point in the sky) the energy available from the sun is roughly 1,000 Watts/meter^2—which means that a solar module that was 100% efficient would produce 1,000 Watts (or 1 kilowatt) for every

square meter of surface. Sadly, there are no perfectly efficient solar modules available. Commercially available units typically have module efficiencies that range from 11% to 20%.

Efficiency is a very important attribute if your available space for solar is limited. However, efficiency is expensive—if you have lots of roof space for solar, the value of this factor drops dramatically.

2. ***Temperature performance***—Solar modules are exposed to the full heat of the sun and as a result, they get very hot. If mounted close to a roof, they get hotter still. Perversely, all solar modules lose efficiency as they heat up, some more so than others. One way to assess temperature performance is to look at the ratio between the *STC* and *PTC* ratings of the modules under consideration.

   The STC rating of a module—also known as its nameplate rating—is the manufacturer's assessment of the module's performance at so-called **S**tandard **T**est **C**onditions. On the other hand, the module's PTC rating, evaluated under **P**hotovoltaic **T**est **C**onditions, is based on more "real world" conditions and is generally accepted as a better predictor of how the module is likely to perform on your roof. (Those real world considerations explain why rebate predictions are based on PTC ratings.)

PTC ratings are always lower than STC ratings for the same module, but the closer the ratio of PTC/STC is to 100%, the better the module will perform at temperature. Those ratios can run from a low of 80% to as high as 94%, but again, better temperature performance comes with a price.

3. ***Manufacturing tolerance***—Since solar modules are manufactured products, there are variances from one module to the next. Manufacturing tolerance refers to how closely the actual power initially produced by the module corresponds to its nameplate rating. Lower quality modules typically exhibit wide tolerances of as much as plus or minus 10%. That means that a nominally 200 Watt module might actually be producing as much as 220 Watts—or, more likely, as little 180 Watts.

   Such wide ranges of performance negatively affect overall system performance, since central or string inverters cannot accommodate such wide swings in module performance. (Microinverters, given that they are combined with only one solar module, can help here since this type of module mismatch is no longer an issue.)

   Higher quality modules have much narrower tolerances, and the best modules will have *positive-only* tolerances. For example, NeoN modules from LG Electronics feature 0 to +3% tolerances—which means you are guaranteed of getting the performance you paid for and excellent overall system performance.

4. ***Manufacturer reliability***—Pretty much all solar modules come with a 5 to 10-year warranty on workmanship and a 25-year warranty on performance, the latter of which purports to tell you how the module will perform over time. While it is termed a warranty, in reality it is more of an aspirational document since, as a practical matter, there is no way to make a claim against the "warranty". Most performance warranties assert that the module will produce 80% of the nameplate rating after 25 years, which works out to a degradation in module performance of roughly 0.9%/year.

   Of course, the value of any warranty is entirely dependent on the reliability of the manufacturer. Lots of cheap solar modules from start-up operations in Asia are now on the market. If you have a problem five years from now and require a warranty replacement, how likely is it that the manufacturer will still be in business? A 25-year warranty issued in 2013 is of no value at all in five years if the company disappears in 2015. Ask your potential installer about the history of the company that is standing behind that warranty, and then do your own homework on this vital reliability factor. Be prepared to pay more for solar modules coming from the most reputable manufacturers.

Inverters

Solar power inverters for commercial installations range from large, *central inverters,* to a collection of *string inverters,* to *microinverters* (one inverter per solar module).[7] You should check for:

a) The efficiency of the inverter (should be in the 95-97% range with the higher the number the better);

b) Whether monitoring is built into the inverter or must be added, and if the latter, how;

c) The warranty period applicable to the inverter; and

d) The manufacturer's reliability. (Inverter recalls in the solar industry are rare, but they have been known to happen.)

There are trade-offs associated with the different inverter options, for example, a central inverter consolidates your equipment in one place and makes for a clean, cost-effective and efficient system, often with sophisticated monitoring capabilities built-in.

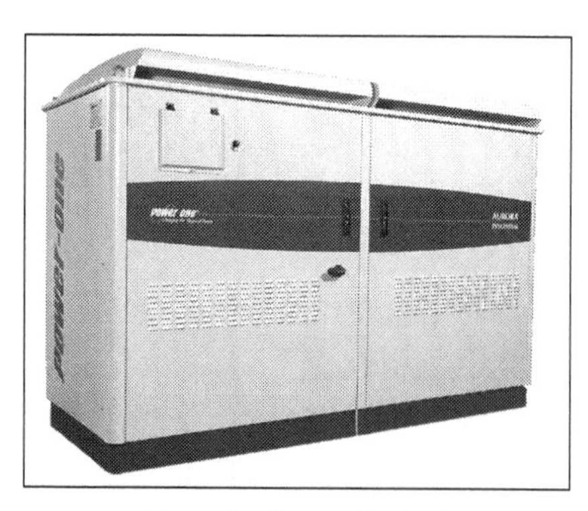

Figure 2 - Aurora Central Inverter from Power-One

However, a central inverter represents a single point of failure for your entire system—if the central inverter goes down, your system will produce nothing until it can be repaired. If the system isn't being monitored on a regular basis, the failure could go unnoticed for an extended period of time, possibly resulting in lost revenue or higher utility bills.

In contrast, using a series of smaller string inverters may look more cluttered, and interconnecting them for monitoring purposes may be more complicated and costly. However, by distributing the inverter function over a number of devices, you have also distributed your risk—if one inverter fails, the others are unaffected and you will continue to produce the majority of your available energy while the faulty device is repaired or replaced. Most commonly, small commercial systems (those below 50 kW) may well benefit from using multiple smaller inverters. As system sizes increase, however, the cost-savings and ease of installation of the central inverter may make it the preferred approach.

Figure 3 - Commercial String Inverters

Courtesy SMA Solar Technology AG

Microinverters have been making inroads into the small commercial market because of better warranties (as long as twenty-five years compared to ten for string or central inverters), higher energy yields (due to shade tolerance, the ability to make use of multiple sub-array alignments, and no problems with module mismatch), and system monitoring down to the module level.

Inverter configuration is ultimately a design choice, and your potential solar contractor should be able to explain to you why they have made the choice that they are recommending.

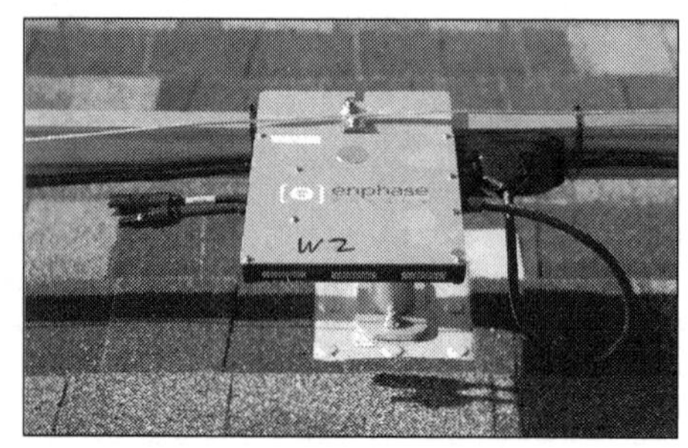

Figure 4 - M215 Microinverter from Enphase Energy

Warranties

We already touched on the value of warranties from equipment manufacturers, but what about the warranty from your solar contractor? In California, all solar companies are required to offer a 10-year warranty on their workmanship. However, a company that has only been in the solar business for a couple of years (or less!) cannot offer proper assurances that they will be there to back-up that warranty. Here's one hint: if your potential solar contractor was in the business before the Great Recession hit, they are probably in this for the long haul (and they have demonstrated enough business savvy to survive the worst economic climate in more than fifty years—not a bad credential).

Rebate Calculations

Your potential solar contractor should have done a calculation based on your utility's rebate structure to estimate what your rebate will be. These estimates should be comparable from one bid to another, but if they are not, demand that your potential contractors provide you with the output from the rebate calculator that they used to produce the estimate. If they refuse, or if the rebate calculation shown doesn't square with the rest of their bid, you can scratch them off your list of potential candidates.

ARE YOU COMFORTABLE?

You cannot be completely comfortable making such an important decision until you have had all of your questions answered. Your potential solar contractor should be happy to spend whatever time you reasonably need to be assured that you have all of the information in hand. The following are the analyses that you should insist on receiving, and having explained to you, before you make your decision.

Utility Savings Analysis

You want to know what your savings will be from your new solar power system, and this analysis should answer that question. A proper Utility Savings Analysis must do three things:

- Predict the production of the system, both in terms of peak power and energy over time;
- Assess the value of that production in Year 1; and
- Apply appropriate factors to assess the change in value of that production over the lifetime of the system.

Let's break this down, starting with your predicted power and energy production.

Power and Energy Production

The peak power and energy yield from the proposed system in Year 1 is a function of system and environmental factors. The system factors include the modules and inverters chosen (including all of the variables discussed previously).

The environmental factors consist of the azimuth (orientation relative to North) and pitch of the array as well as any shading factors that might be present. If the overall array is comprised of sub-arrays with different environmental factors, then each sub-array must be assessed separately.

For a so-called "fixed-plate array"—that is a solar array that is at a set azimuth and pitch, which is typical for commercial installations—the ideal azimuth is due South and the ideal pitch is the latitude of the site. While deviations from these ideal values will result in reduced annual energy yield, in the real world such deviations are common. Indeed, in some settings a deviation might be desirable if, for example, summertime performance is to be maximized (perhaps to mesh with the a time-of use rate structure or the payment profile of a feed-in tariff program[8]), in which case a flatter array pointed more to the west might be selected.

Regardless of the azimuth and pitch, shading is to be avoided, especially if string or central inverters are used. When a string of solar modules are wired together, shade falling on one module not only degrades the performance of that module, it will degrade the performance of the entire string. This in turn will degrade the performance of the entire sub-array of which that string is a part. Microinverters overcome this problem because each module operates independently—a shaded module still sees its performance deteriorate, but that deterioration has no effect on the adjacent, unshaded modules.

All of these factors, as well as the geographic location of the system site, are then provided as inputs into a PV system performance model, the best known being *PVWatts II*, created by the National Renewable Energy Laboratory (NREL). PVWatts is used as the underlying analysis engine of many utility rebate calculators, including the CSI rebate calculator. The output from the calculator will provide a value for the peak output from the system (in AC Watts) and the energy yield profile over a year—either month-by-month, or even hour-by-hour.

Savings in Year 1

Knowing the production profile for the proposed system is just the first step in the utility savings analysis. The next crucial step is to calculate the savings from that production in Year 1—the first year the system goes live. To do this accurately requires a detailed analysis of the relevant utility rate schedule and possibly detailed information about how the existing loads at the site behave. A simple-minded analysis that assumes that all kWh's of energy are worth the same fails to meet this standard and will not accurately predict the savings to be derived.

As noted before, most commercial solar customers pay for both total energy usage and peak power demand. To determine your anticipated savings accurately requires a clear understanding of how the solar power system will affect both of those components. Unfortunately, it is not uncommon that the data necessary for such an analysis will be incomplete or missing altogether.

Usage savings are easy to calculate. If you have past usage data (generally available from the utility except when the potential client is in a new building), and a properly designed rate structure model, it is easy to apply the energy yield profile from the performance calculator to the rate structure and determine savings. For customers on usage only rate structures, this provides a nearly perfect estimate of annual savings as of Year 1.

Things get complicated when trying to determine how the solar power system will alter demand charges. To do this accurately—and honestly—requires hour-by-hour demand data so that the client and the solar contractor know when peak demands occur. If the peak demand occurs at solar Noon, the solar system will directly reduce that peak - perhaps by the full value of the solar power system. Conversely, if the peak demand occurs at eight o'clock in the morning - when the first shift arrives - the solar power system will have next to zero effect on peak demand.

Unfortunately, unless the potential client has been on a time-of-use based rate structure, such demand data is almost certainly unavailable.

Under those circumstances client and contractor have only two choices: gather the data as part of the site evaluation process (by temporarily or permanently installing data logging equipment), or make a well-documented estimate of what the effect of the solar power system will be on peak demand. If that is the path chosen, the client should insist that all of its bidders use the same estimate.

Fortunately, relatively inexpensive data-logging equipment is now available, and it should be used whenever the decision-maker's timing permits. A contractor who refuses to provide such a service—for a fee, of course—should be scratched from the list of potential candidates.

Savings Over System Lifetime

So now you have an estimate of your utility savings in Year 1—how can you determine what your savings will be over the lifetime of the system? After all, that is the key question in determining your ultimate Return on Investment (ROI).

To answer that question requires figuring out two more puzzle pieces—one straight-forward and the other remarkably controversial. The straight-forward puzzle piece is how will the system's performance change over time. This is straight-forward because we have reliable data for making that prediction. Assuming reasonable maintenance for the system—cleaning the modules occasionally, replacing or repairing broken modules or inverters, *etc.*—the performance from one year to the next is really just a function of the deterioration of the installed solar modules' performance. That rate will be documented in the performance warranty of the module, and hence it can be easily modeled. (As noted previously, for most modules with a performance warranty that guarantee 80% of nameplate power after twenty-five years, that works out to a degradation factor of -0.9%/year.)

The controversial puzzle piece is in "guesstimating" what will happen with utility rates over the lifetime of the system. Not only is this a difficult task at best, it has even led to class-action litigation when solar leasing giant Sunrun was sued for over-estimating (according to the plaintiff) the magnitude of utility rate increases in the future.

Over the years solar companies have used annual rate increase factors ranging from 6.8% (almost certainly too high) to 3% (almost certainly too low, at least in California). Sunrun was sued for picking 6%, and yet in 2012 SCE secured a three-year, 17.2% average rate increase—which works out to 5.7%/year!

So what is the right number to use? At Run on Sun we generally use 4.5% for municipal utilities and 5.7% for SCE. But in our view, ***the rate selected is not as important as the need to disclose, in a clear and well documented manner, the rate being used in the model***. And of course, the client is free to insist that all potential solar contractors bidding on the project use the same rate.

However the rate is determined, and ultimately disclosed, the result should be a series of values for how much savings will be generated by the system for the next twenty-five years—and that constitutes your Utility Savings Analysis. Given that result, your potential solar contractor is now in a position to provide you with your second key analysis, demonstrating your anticipated Return on Investment.

Return on Investment

Given the energy saving starting in Year 1, the cost of the system, any O&M costs, the anticipated rebate from the utility, and the tax benefits anticipated for the system, your installer should map out for you the cash flows associated with your system.

That analysis should indicate when the system will break even, and what the internal rate of return (IRR) will be over the lifetime of the project. A competent analysis will also identify all of the assumptions used in the analysis.

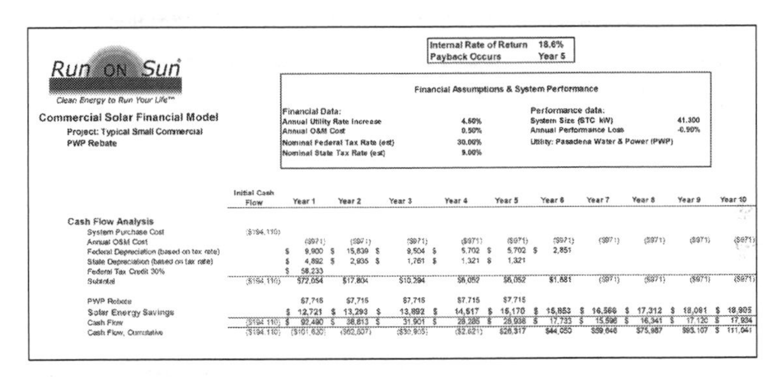

Run ON Sun
Clean Energy to Run Your Life™

Commercial Solar Financial Model
Project: Typical Small Commercial
PWP Rebate

Internal Rate of Return	18.6%
Payback Occurs	Year 5

Financial Assumptions & System Performance

Financial Data:		Performance data:	
Annual Utility Rate Increase	4.60%	System Size (STC kW)	41.300
Annual O&M Cost	0.50%	Annual Performance Loss	-0.90%
Nominal Federal Tax Rate (est)	30.00%	Utility: Pasadena Water & Power (PWP)	
Nominal State Tax Rate (est)	9.00%		

Cash Flow Analysis	Initial Cash Flow	Year 1	Year 2	Year 3	Year 4	Year 5	Year 6	Year 7	Year 8	Year 9	Year 10
System Purchase Cost	($194,110)										
Annual O&M Cost		($971)	($971)	($971)	($971)	($971)	($971)	($971)	($971)	($971)	($971)
Federal Depreciation (based on tax rate)		$ 9,900	$ 15,839	$ 9,504	$ 5,702	$ 5,702	$ 2,851				
State Depreciation (based on tax rate)		$ 4,892	$ 2,935	$ 1,761	$ 1,321	$ 1,321					
Federal Tax Credit 30%		$ 58,233									
Subtotal	($194,110)	$72,054	$17,804	$10,294	$6,052	$6,052	$1,881	($971)	($971)	($971)	($971)
PWP Rebate		$7,716	$7,715	$7,716	$7,715	$7,715					
Solar Energy Savings		$ 12,721	$ 13,293	$ 13,892	$ 14,517	$ 15,170	$ 15,853	$ 16,566	$ 17,312	$ 18,091	$ 18,905
Cash Flow	($194,110)	$ 92,490	$ 38,813	$ 31,901	$ 28,285	$ 28,938	$ 17,733	$ 15,596	$ 16,341	$ 17,120	$ 17,934
Cash Flow, Cumulative	($194,110)	($101,620)	($62,807)	($30,905)	($2,621)	$26,317	$44,050	$59,646	$75,987	$93,107	$ 111,041

Figure 5 - Return on Investment Analysis

While IRR is a somewhat arcane concept—it is defined as the interest rate for which the Present Value of all of the future cash flows associated with a given project is equal to the cost of building the project[9]—its real benefit is in comparing competing investments, or solar power system proposals in this case. As with any earned interest value, the higher the IRR the better the investment.

Levelized Cost of Energy (LCoE)

While it is common in the solar industry to express the cost of the system in dollars/Watt, that is a misleading statistic at best since it masks variables affecting real world performance. A far better metric—and one that your installer should be able to provide you—is the cost per kWh for the energy that will be produced by the system over its anticipated lifetime.

The calculation is actually quite simple: determine the total out-of-pocket costs for the system owner over the system's lifetime (consisting of the purchase price *reduced* by the rebate and tax credits, *increased* by all O&M costs) and divide it by the total amount of energy to be produced (allowing for the system's performance degradation over time) over the system's lifetime of twenty-five years.

We prefer this number because it reflects the real world performance, and it allows for direct comparisons against the client's previous costs for energy. Indeed, we typically find costs per kWh in the 6-10¢ range for solar power systems compared to utility costs of 15-25¢ starting in Year 1. But because the energy cost for the solar power system is fixed over its entire lifetime while the cost of energy from the utility is constantly rising (even if we don't know precisely how fast), the comparison is quite compelling.

Note that by applying an agreed upon (or at least disclosed) rate for utility increases, a graphical comparison over time can be produced—but the underlying LCoE is not at all dependent upon future utility rate changes. This gives the client the ability to compare multiple proposals against a true value proposition—how much will the energy from the proposed system cost? From a financial perspective, this is the best comparison point that we have been able to identify. A potential solar contractor who balks at providing this should, you guessed it, be scratched from your list!

Chapter Notes

[1] Visit the NABCEP website to learn more about their accreditation process and to access their database of Certified PV Installation Professionals, http://www.nabcep.org/, accessed 6/2/2013.

[2] IBEW is the International Brotherhood of Electrical Workers, http://ibew.org/, accessed 6/16/2013.

[3] NABCEP Certification brochure, http://www.nabcep.org/wp-content/uploads/2011/03/NABCEP-Consumer-brochure.pdf, accessed 7/3/2013.

[4] The *Go Solar California* installers database can be found at: http://www.gosolarcalifornia.ca.gov/database/search-new.php, accessed 8/20/2013.

[5] An August 2013 analysis of CSI data revealed that the largest solar installers averaged more than five months to install a solar power system, see, *Outliers and Oddities – 2013,* http://runonsun.com/~runons5/blogs/blog1.php/stsolar/csi2013/outliers-oddities-state-of-socalsolar-2013, accessed 9/2/2013.

[6] That same study of CSI data revealed that choosing larger installers to do a project was no guarantee of receiving the most favorable price.

[7] While microinverters are certainly not the rule for very large commercial systems, we are aware of a 2.3 MW system that was installed using Enphase microinverters. See Enphase Energy blog, http://enphase.com/eblog/2013/featured-array-vine-fresh-produce/, accessed 8/20/2013.

[8] More information about feed-in tariff programs is in Chapter VI. Special Cases: Feed-in Tariffs, p. 97.

[9] Definition of IRR from the American Institute of CPAs' website, http://www.aicpa.org/InterestAreas/ForensicAndValuation/Membership/DownloadableDocuments/Intl%20Glossary%20of%20BV%20Terms.pdf, accessed 8/20/2013.

IV. FINANCING

JACK AND THE MYSTERIOUS WORLD OF PROJECT FINANCE

From the back of the cab, Jack hung up his cellphone and permitted himself a satisfied smile. Just weeks ago he had gone from knowing nothing about solar power to having found three qualified contractors who had each given him legitimate proposals. He had scored each one and had been able to beat his CFO's deadline for bringing a recommendation to the CEO in time for the annual meeting.

"NEARLY perfect" *indeed*, he chuckled to himself.

The Board had been pleased with the CEO's vision—which meant that the CEO was pleased with Jack's execution. Now only one thing remained before the project—a 55 kW system on ExGen's main building—could begin.

Figuring out how to pay for it.

Certainly one option was for ExGen to pay cash. After all, the company was doing well, and they could shift some things around to cover the $235,000 price tag of the preferred proposal. But ExGen's CFO was pretty tight with the company's cash, and Jack knew that she would at least want Jack to explore some other options.

And that's where Jack's lunch date fit in. Pulling up to the All Greene Cafe—a resolutely hip vegetarian restaurant that Jack had only previously read about—he realized how much he was looking forward to meeting Joy.

Joy Cern was a habitué of AGC—as the hipsters referred to it—and she had eagerly recommended it to Jack for their lunch. More importantly to today's meeting, Joy was also a finance whiz who specialized in the various ways of paying for renewable energy projects, making her just what the doctor had ordered.

After a great meal of dilled asparagus soufflé—which prompted Jack to make a mental note to invest in a more sophisticated vegetarian cookbook—the conversation got down to business.

"There are a number of interesting options available these days, Jack, so you shouldn't need to spend your cash unless you want to," Joy insisted. "For example, there are loans, leases, PPAs and PACE. Where would you like to start?"

"I think I understand about loans," Jack said, "and I've heard about leases for solar. But I have no idea what those other two are."

"Okay," said Joy. "But let's start with leases since there are at least two kinds of leases—a capital lease and an operating lease—and they are very different.

"With a capital lease, you will own the equipment, and you will get the rebate and the tax incentives. That also allows you to claim credit for having a solar power system on your building without running afoul of the Federal Trade Commission's regulations.

"An operating lease, on the other hand, might have a lower monthly payment, but you won't own the equipment, and the leasing company takes the rebate and tax incentives.

"However, if the size of the project is small—typically under $250,000—the leasing companies may not be interested."

"Oh," said Jack, "that might be a problem for us since our project size is slightly smaller than that."

"It doesn't hurt to ask," Joy replied. "Depending on the leasing company they might be willing to bend their rules a bit for a particular project.

"Another option is the Power Purchase Agreement, or PPA, which is like an operating lease, but instead of a fixed monthly payment that is based on the value of the equipment, with a PPA you only pay for the energy that the system actually produces.

"Keep in mind, though, that both leases and PPAs might have what are sometimes called accelerator or escalator clauses—which allows the monthly payment to increase over time. If that rate goes up faster than the utility's rates do, you might not get the savings that you were hoping for."

"Right," said Jack, "that makes sense. So what is this last option?"

"That would be PACE," replied Joy, "which I think is the most interesting arrangement of all. PACE stands for Property Assessed Clean Energy, and it originally began as a program for residential projects in San Francisco. Although residential PACE programs are mostly stalled, they are very much available for commercial projects like yours.

"With a PACE program—such as the one that we have here in Los Angeles County—you pay for the solar power system as a special assessment on your property tax bill. You own the system so you get the rebate and tax benefits. But if you ever sell the building, the new owner assumes the payments as part of their property taxes. PACE programs generally have easier qualification terms since the debt is tied to the property, not the company purchasing the system. Plus, most PACE programs for commercial solar projects will fund smaller projects. For example, the LA County program will fund projects as small as $150,000.

"The only downside to PACE is that it is not available everywhere, and you have to have a property tax bill to participate—which can exclude some non-profits."

"So how do I figure out which one will be best for us?" asked Jack.

"Your preferred contractor should be able to help you get started in terms of applications. Ultimately, you will want to run the numbers with your CFO to determine which option is best for EnGex," said Joy.

Installing a solar power system is a substantial investment, and part of what determines your return on that investment is how the system is financed. In recent years a great deal of creativity (some would say perhaps too much creativity) has been brought to bear on the subject of how to finance solar systems, resulting in the introduction of myriad financing schemes from the terribly simple (straight cash purchase) to the terribly complex (e.g., flips and swaps). As the amount of money in play increases, the more complex the schemes become.

Fortunately or unfortunately, in the realm of small to mid-sized commercial solar systems, the options are more limited and include cash purchases, loans, various types of leases, and Power Purchase Agreements (PPAs). Let's look at each of these in turn.

CASH OR SELF-FINANCED PURCHASES

As Joy explained to Jack, the simplest financing method is the cash purchase—simple, that is, if you have the cash on hand and it isn't needed elsewhere.

When a company self-finances through a cash purchase, they own the system outright and receive the rebate payment from the utility and all of the tax benefits. For those entities with the cash on hand, a cash purchase may be the best possible option since, unlike all of the other methods available, there is no added cost to the price of the system. Instead, a solar power system that is purchased outright should be looked at in terms of its opportunity cost. That is, what advantage/disadvantage does the solar investment provide compared to where the same capital could have otherwise been invested?

These days, with interest rates at historic lows, capital invested in traditional savings instruments, such as savings accounts or certificates of deposit (CDs), provide safety, but returns in the 1-2% range which are not terribly attractive. On the other hand, investments with higher returns, such as individual stocks or stock funds, come with substantial risk, as the crash of 2008 painfully reminded us.

As a result, a solar power system—with next to no risk and an IRR of 12-19%—compares quite favorably. Put most simply, a safer investment—if you could find one—would provide a far, far lower return whereas an investment with a higher yield would be far, far riskier.

When viewed through such a lens, a solar power system becomes a very attractive investment indeed. In fact, when analyzed in that fashion, investing in solar even makes sense as a way of employing endowment funds designated for the maintenance of non-profit organizations like private schools and churches.

LOANS

Unfortunately, not every entity that would like to add solar is in a position to self-finance. For those who must seek other financing sources, a conventional loan is the obvious alternative—if it is available. Although interest rates remain at near historic lows, many banks are historically reluctant to make loans at all, let alone for "exotic" projects like solar power installations. Or, if they are willing to consider it at all, they may impose onerous terms or prohibitively restrictive conditions that keep solar loans more of a theoretical option than a practical one.

Bankers are focused on collateral and cash flow considerations, with solar being strong on the latter but notoriously weak on the former. Normally a loan for an equipment acquisition could be collateralized by the equipment itself—if you don't pay on your car loan, the bank repossesses the car. But repossessing a solar power system is a complicated prospect, and unlike a used car which has a readily determined resale value, the resale value of used solar equipment is uncertain, at best.

On the other hand, solar power systems significantly enhance the cash flow situation of the loan customer since the sum of the remnant electric bill plus the loan payment should be substantially less than the old electric bill. Moreover, with a fixed interest rate loan, that difference will only improve over time.

In the end, the availability of loans for these projects comes down to a question of the banker's comfort level with solar. Does the reduced risk of the customer defaulting, thanks to the improved cash-flow prospects, outweigh the downside of increased risk due to poor or no collateral?

LEASING

One option that has gained significant traction in the past few years is the solar lease. Indeed, it is the explosion of solar leasing in the residential market that has fueled the growth of major players like SolarCity and Sunrun. But leases can come with unexpected traps for the unwary. A commercial customer needs to look closely at the details before signing on to a lease agreement for a commercial solar power system.

In a solar lease arrangement the right to use the solar power system is transferred from the owner, referred to as the lessor, to the lessee. From an accounting perspective, all leases are either considered a capital lease (*aka* a finance lease) or an operating lease. Generally speaking, "capital leases are considered equivalent to a purchase, while operating leases cover the use of an asset for a period of time."[1]

When leases are applied to solar power systems, other important considerations apply.

Capital Leases

Under accounting standards, a capital lease is defined as "a lease that transfers substantially all the benefits and risks of ownership to the lessee."[2]

Therefore, with a capital lease, as with a cash purchase or loan, the solar client is treated as the owner of the system and receives the benefits of ownership: utility rebates and tax incentives (if applicable). The capital lease is often for a longer term—the basic criteria is that the lease must run for at least 75 percent of the estimated economic life of the system—that is, 15 years or longer.

The longer term can keep payments lower, but because the solar client lessee receives the rebate and tax incentives, the capital lease might carry a higher interest rate than does an operating lease. At the end of the term, the lessee can typically purchase the system at below market cost, perhaps for as little as a nominal one dollar.

Operating Leases

In contrast, with an operating lease the solar client lessee does not effectively own the system and the lessor retains the utility rebate and the tax incentives. Rather, the lessee is simply acquiring the right to use the system for a limited time in exchange for periodic rental payments. Typically, an operating lease will be for a shorter period of time, and potentially at a lower interest rate. However, at the end of the lease term the lessee either has the system removed by the lessor, enters into a new lease arrangement, or must purchase the system for fair market value.

Power Purchase Agreements

A related, but different, vehicle for financing a solar power system is a Power Purchase Agreement, or PPA. As with an operating lease, the solar client under a PPA does not own the system. Rather, they purchase the energy that the system produces from the system owner, presumably at a price lower than what they would be paying their utility for the same quantity of energy.

Since the solar client under a PPA only pays for the energy actually produced by the system, the system owner has a greater incentive to maintain the system at peak efficiency and the solar client may receive more of the "benefit of their bargain" under a PPA than they would under an operating lease.

PPAs typically contain "escalator clauses" by which the price paid per kilowatt hour generated may increase over time. As long as the cost of energy under the PPA increases slower than the corresponding utility's energy rates, the solar client's savings will grow over time. However, it is possible under a PPA to end up paying more for energy to the system owner than the client would have paid to the local utility. (Indeed, this possibility is what gave rise to the class action lawsuit against Sunrun.[3])

Federal Trade Commission Concerns

Being locked into a potentially bad deal is not the only possible concern for the lease or PPA customer—they might also run afoul of federal law. The Federal Trade Commission (FTC) is the federal agency charged with regulating false or deceptive marketing claims, and solar leasing options can surprisingly lead to unwanted scrutiny from the FTC. The FTC's concern is "double counting"—multiple entities taking credit for the same environmental benefit. This sort of double counting can occur when a company hosts a solar power system, but does not own it.[4]

The FTC provides the following as an illustrative example:

> A toy manufacturer places solar panels on the roof of its plant to generate power, and advertises that its plant is "100% solar-powered." The manufacturer, however, sells renewable energy certificates [RECs] based on the renewable attributes of all the power it generates. *Even if the manufacturer uses the electricity generated by the solar panels, it has, by selling renewable energy certificates, transferred the right to characterize that electricity as renewable.* **The manufacturer's claim is therefore deceptive**. *It also would be deceptive for this manufacturer to advertise that it "hosts" a renewable power facility because reasonable consumers likely interpret this claim to mean that the manufacturer uses renewable energy*. It would not be deceptive, however, for the manufacturer to advertise, "We generate renewable energy, but sell all of it to others."[5]

Deceptive claims are actionable under the FTC's mandate, and offending companies could be subject to enforcement actions and fines. Under either an operating lease or a PPA (though likely not under a capital lease unless the RECs are surrendered to the utility), the solar client does not own the solar power system, and any claim to be "solar-powered" or "using green energy" would be deceptive under the FTC's guidance.

COMMERCIAL PACE

Another option that is starting to appear is PACE financing. PACE—which stands for Property Assessed Clean Energy—operates in cooperation with a local government, typically a city or county, that agrees to finance solar power systems through the sale of municipal bonds. Investors purchase the bonds, and the proceeds are used to pay for the installation of the solar power system. The government entity imposes a lien on the property to be discharged after the owner has paid back the system cost over time as an assessment on their annual property tax bills. If the client chooses to sell the property, the obligation "runs with the land" and is assumed by the new owner (who, of course, also derives the benefit from the solar power system).

Under PACE, there is no personal obligation on behalf of the solar client, so neither corporate nor personal credit is at issue. In theory, PACE has the potential to increase greatly the number of entities that could qualify for solar financing.

Unfortunately, to date, PACE has yet to live up to that potential. Jurisdictions have been slow to institute PACE programs, and even in cities and counties where it has been adopted—such as in Los Angeles County—the pace of PACE-funded projects has been discouragingly slow. According to the New York Times, the total number of PACE-funded projects throughout the entire United States is less than two hundred (worth less than $40 million), although that number is expected to double in the next year.[6] Part of the problem is the reluctance of some investors to comprehend the benefits of PACE as an investment vehicle, and the (perceived, if not real) inability to resell PACE investments in the secondary market. Until those issues are resolved, PACE will remain more promise than reality.

CROWD FUNDING

The latest trend to hit solar financing is that offered by Crowd Funding companies like Solar Mosaic[7] which provide an online platform intended to bring together individual investors with selected solar projects. At the Solar Mosaic website, potential investors can review projects and invest however much they choose, in $25 increments. However, investors must be "qualified" based on income and/or net worth (without counting autos or residence). Investors who do not satisfy the qualification criteria have their total investment in any twelve months capped at $2,500.[8]

By the end of June, 2013, Solar Mosaic had reported funding fourteen projects worth a combined investment of $2.1 million. The loans being provided by Solar Mosaic, however, are not covering the full cost of the systems being built. Rather, they appear to be limited to something on the order of 25% of the total project cost.

Solar Mosaic offers an innovative, if limited, model for solar project financing. It will be interesting to see if the crowd funding model succeeds with solar and expands over time, as well as whether competitors, like UVest Solar,[9] can build on what Solar Mosaic started.

FINANCING LIMITATIONS

Regardless of the financing vehicle—other than cash purchases—there are some common limitations as to the applicability of any of these methods. The most common impediment is the size of the project. Because all of these financing methods involve some amount of overhead, small projects are typically harder to fund, with project thresholds of $150,000 or even $250,000 being common. While these limits aren't a problem for mid-sized commercial projects, they can, in effect, leave small commercial projects in the 30-60kW range unfunded.

The credit worthiness of the solar client is also a consideration for each of these methods, except PACE. Non-profit organizations might find themselves shut out of all of these funding methods because of concerns for longevity, in some cases, or simply because they do not pay property tax bills (a deal breaker for PACE programs).

Since non-profits do not qualify for tax benefits, their cash flow improvement is not as great as it is with their for-profit neighbors. Moreover, whereas a small commercial customer might be able to secure a loan by making a personal guarantee (indeed, that may well be required), a non-profit organization is unlikely to have anyone qualified (or inclined) to provide such a guarantee.

More creative approaches may be needed for non-profits. For example, some non-profits are fortunate enough to have endowment funds that are restricted in how they can be used, but which might exceed many times over the cost of the proposed system. Diverting some of those funds into a separate, interest-earning account against which the lending institution can attach a lien provides adequate collateral for the lender, with possibly acceptable risk to the endowed funds.

Non-profits that are not so well endowed, but which have a well-established donor base, could consider the possibility of creating a free-standing, for-profit corporation to own the solar power system and to provide a PPA back to the system-hosting non-profit. Since the for-profit owning entity can secure tax benefits, it can make the venture financially viable even if more conventional funding cannot be found.[10]

Chapter Notes

[1] *Capital and Operating Leases A Research Report*, Susan S.K. Lee, Federal Accounting Standards Advisory Board, October 2003, p. 9, http://www.fasab.gov/pdffiles/combinedleasev4.pdf, accessed 5/26/2013.

[2] Lee, p. 10.

[3] *Solar lease companies face criticism over calculating energy savings*, Felicity Carus, March 7, 2013, http://www.pv-tech.org/news/solar_lease_companies_face_criticism_over_calcluating_energy_savings, accessed 5/26/2013.

[4] Problems with the FTC can also arise if the renewable energy credits associated with the project were sold to the local utility as part of the rebate payment.

[5] *Guides for the Use of Environmental Marketing Claims ("Green Guides")*, Federal Trade Commission, October 1, 2012, http://www.ftc.gov/os/2012/10/greenguides.pdf, accessed 5/26/2013.

[6] *Tax Programs to Finance Clean Energy Catch On*, by Diane Cardwell, New York Times, 6/21/2013, http://www.nytimes.com/2013/06/22/business/energy-environment/tax-programs-to-finance-clean-energy-catch-on.html, accessed 6/24/2013.

[7] Solar Mosaic website, https://joinmosaic.com/, accessed 5/27/2013.

[8] Investor qualifications are based on rules from the federal Securities Exchange Commission (SEC) and the California Corporations Commissioner. http://support.joinmosaic.com/entries/21981786-Who-can-invest-through-Mosaic-s-platform-, accessed 5/27/2013.

Chapter Notes, *continued*

[9] UVest Solar website: http://www.uvestsolar.com/, accessed 5/27/2013.

[10] Of course, such an enterprise must be carefully crafted under the guidance of an experienced tax attorney.

V. YOU'VE SIGNED A CONTRACT, NOW WHAT?

JACK PRINCE: PROJECT MANAGER

Jack arrived at EnGex with more spring in his step than usual, filled with the confidence that this was going to be an exceptionally good day. After months of research, comparing proposals, investigating financing options—and dealing with the always challenging internal politics of EnGex—Jack was finally signing a contract with his preferred solar contractor, Run on Sun. As rigorous as the process had been, Jack knew that the real work was just beginning and he was eager to get underway.

So it was with real pleasure that Jack welcomed Brad back to his office.

"Great to see you again, Brad. Congratulations on being selected."

"Thanks, Jack," said Brad, "we are always happiest when the client has a detailed and thorough process like yours. It means that they have really done their homework, and that always results in an easier implementation."

"That's really what I wanted to talk about with you today," replied Jack. "As you know, this is my first solar project and I really don't know what to expect. Can you walk me through what your process will be?"

"Absolutely," said Brad, "first things first. In addition to signing our contract today, I have also brought with me the paperwork that your utility requires us to submit to secure your rebate reservation. It is important that we get that submitted immediately. That way, if the utility lowers its rebate rates, your rebate will already be locked in."

"Ok," said Jack, "do we have to handle that or will you?"

"No," replied Brad, "Run on Sun will handle everything, you just have to sign the papers."

"And the checks," interjected Jack.

"True enough," agreed Brad. "There's no such thing as a free lunch!"

"I'm just giving you a hard time," confessed Jack. "Sorry to interrupt. So once the rebate application is submitted, then what happens?"

"We wait. Your contract has a contingency that if for some reason the rebate is not approved, you can terminate the contract, and that is for your protection. But we have never had a rebate application rejected, so that is highly unlikely to happen now.

"Once the utility advises us that the rebate has been reserved, we will assemble the paperwork needed to pull the permits for this job. In your jurisdiction that means a building permit—for the structural aspects of the job—and an electrical permit. We will prepare detailed drawings of the site and proposed installation, as well as an electrical layout, called a single-line drawing.

"To support our permit application we will submit wind and seismic load calculations from a structural engineer, and we will have an electrical engineer sign off on our single-line drawing. Put all that together, along with a hefty fee, and we should then be able to pull our permit and schedule the job."

"How long will all that take?" asked Jack.

"Unfortunately, that is really hard to say," admitted Brad. "Some jurisdictions will do this over the counter and, if they don't require any changes, you can walk out of the office with a permit the same day. Others require you to submit every application to Plan Review which means they will sit on it for weeks before they will even look at it. In those cases it can take many weeks to pull the permit.

"Fortunately, the process is pretty clean here in Pasadena, so we would expect to have a permit within a couple of weeks."

"That's great. So then what?"

"Well, a project like this really has two phases: one on the ground and one on the roof. They will overlap to some degree, but it is often easier to think of them as distinct.

"On the ground we will install all of the components that are needed to interconnect your system to the grid. Specifically, in this case that will include a good sized transformer, a solar-only subpanel, a couple of disconnect switches and a performance meter. We will also install the networking equipment to support the monitoring system.

"On the roof we will start with what we need to attach the array to the roof structure. If this were a flat roof, we might use a ballasted system which uses just a few physical attachments—for seismic bracing—and concrete blocks to weigh the system down against wind loads. But this is a sloped roof, so we will start with standoffs that are bolted into the roof structure and flashed to prevent leakage.

"On top of the standoffs we will attach the rails and onto the rails we will mount our microinverters and then the solar modules themselves. Once all of that is in place we will wire everything together and, before you know it, we will be done."

"You make it sound easy," said Jack.

"There are always challenges," conceded Brad. "No two roofs are the same, so something unexpected always seems to arise. But it is a process that we have refined over the years, so we are confident that we can handle whatever comes along.

"Once we are done, we will test everything to make sure the system is working as planned. When we are satisfied that it is, we will call for our inspection."

"Ah yes," said Jack, "inspectors, the bane of everyone's existence."

Brad laughed. "They can certainly be, how should I say this, *unpredictable*. But we have never had a real problem with an inspector. We provide detailed plans and then we follow those plans, so there are very few deviations to cause them to question what we have done. Plus, we communicate with our inspector along the way so if they have any questions we can answer them as we go. It creates a more collaborative environment which really pays off when the project is complete."

"That sounds like a very practical approach," said Jack. "How soon after the inspection will we be up and running?"

"Again, that depends," said Jack. "We will notify the utility once we pass the inspection. Then it is just a matter of waiting for them to process the paperwork. That could take a week or two here in Pasadena, but it could be as long as three months in some jurisdictions. Eventually, the utility will issue a

Permission to Operate letter and that is your signal that you can turn the system on and begin generating clean energy for the next twenty-five years!"

"I can't wait," said Jack. "Let's get that contract signed!"

The process Brad outlined to Jack is roughly what you should expect with your project and there really are a predictable set of steps that should be followed, once the contract is signed:

- Prepare and Submit the Rebate Application;
- Pull a Permit;
- Schedule and Perform the Work on the Ground and on the Roof;
- Handle the Inspections; and
- Apply for Permission to Operate and for Rebate(s).

In the sections that follow, we will examine each of these steps from the perspective of an actual installation project that took place at the Westridge School for Girls in Pasadena, California in 2012.

THE REBATE APPLICATION

The rebates being offered from Pasadena Water & Power (PWP) for this non-profit project were scheduled to step-down on December 1, 2011. Indeed, this was such a substantial rebate reduction—a full 26%—that failure to secure the existing rebate rate would have amounted to a loss of tens of thousands of dollars to the client. Moreover, PWP had made it very clear that unless applications were 100% complete and correct, they would be rejected. If that were to happen, a new rebate application would have to be submitted—at the lower

s. Clearly the pressure was on to get this st time!

The application package consisted of eight parts—most of which were straight-forward, but a couple required substantial work to guarantee that the application as submitted would be acceptable the first time. Those parts were:

1) Signed Rebate Application (PWP's form, signed by the client and the contractor under penalty of perjury);
2) Single-Line Diagram (SLD) for the electrical components of the system;
3) Site Plan;
4) CSI Report (as produced by the California Solar Initiative's rebate calculator);
5) Shading Analysis (*i.e.*, a Solar Pathfinder report to determine the shading values used to create the CSI Report);
6) PWP's Net Metering Agreement (executed by the client);
7) PWP's Net Metering Surplus Compensation form (for AB 920 compliance); and
8) Installation Contract between the client and contractor.

In addition, because this was a project for a non-profit client, proof of non-profit status was also required.

Shading Analysis & CSI Report

PWP wisely requires the submission of a shading analysis in addition to the output from the CSI rebate calculator. Since the amount of shading at the site directly impacts the performance of the system—and hence the CSI AC Watts of the system (or the predicted annual energy output in the case of a PBI rebate)—it really doesn't make sense for a utility simply to trust that the installer is telling the truth about shading.

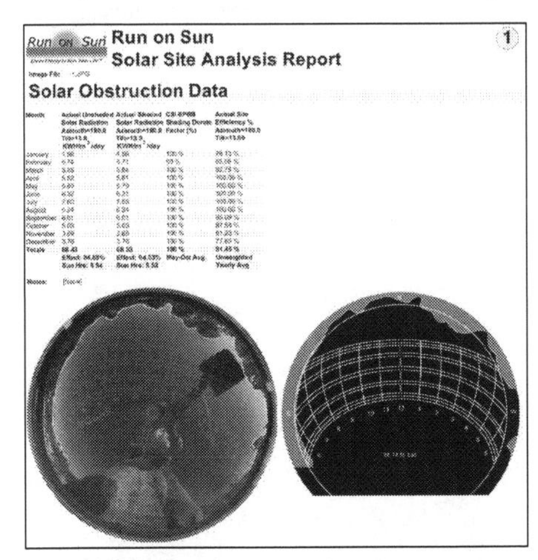

Figure 6 - Solar Pathfinder Shading Analysis

The output from the Solar Pathfinder proves that the shading numbers claimed are the shading values present at the site.

Site Plan

The site plan needed for the rebate application requires an indication of where the various components of the system will be relative to the overall site. However, this particular system was designed to occupy three different areas of the building being used: the roof where the array itself is located, a ground-level storage area where the step-up transformer will be placed, and the utility switchgear, located on the far north end of the building. Thus, the site plan included drawings for each location.

The Site Plan Array Detail drawing (Figure 7) showed the three sub-arrays and the clear space allocated for fire department access. Each sub-array consisted of three branch circuits, each of which was "center-tapped" to reduce the voltage drop in the associated branch circuits. Each branch circuit landed at a sub-array service panel which then fed a master "solar-only" subpanel in the transformer area.

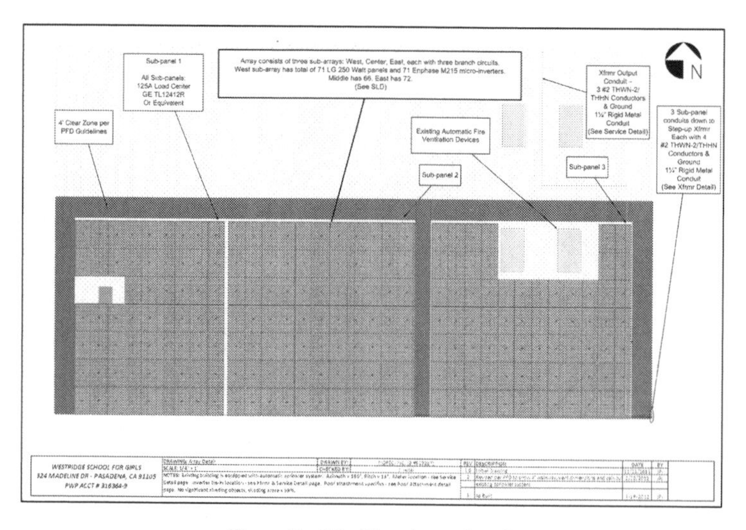

Figure 7 - Site Plan Array Detail

The Transformer Area Detail drawing (Figure 8) detailed the conduits coming down from the roof (one from each sub-array), the master subpanel which feeds our step-up transformer (to change the 208 VAC three-phase power coming from the roof to 480 VAC three-phase supplied by the utility service), and then a safety disconnect switch located adjacent to the transformer. From the safety switch a fourth conduit carries the required conductors back across the roof to the service switchgear area.

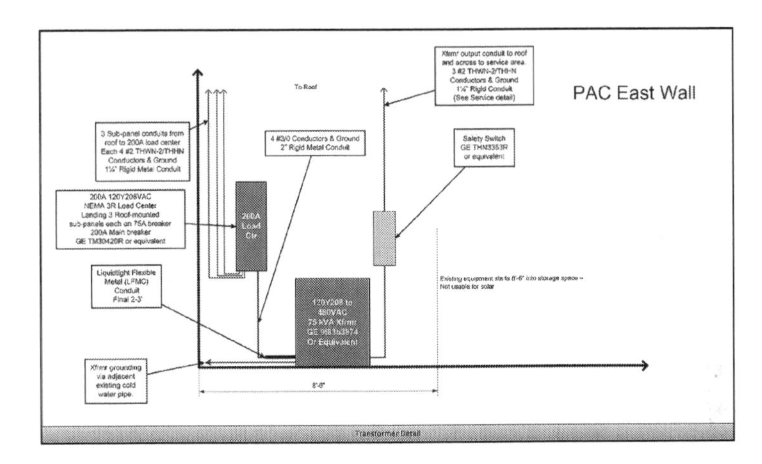

Figure 8 - Transformer Area Detail

The Service Area Detail drawing (Figure 9) showed the placement of the lockable AC Disconnect, the associated performance meter, and the circuit breaker for the system located in the existing service switchgear room.

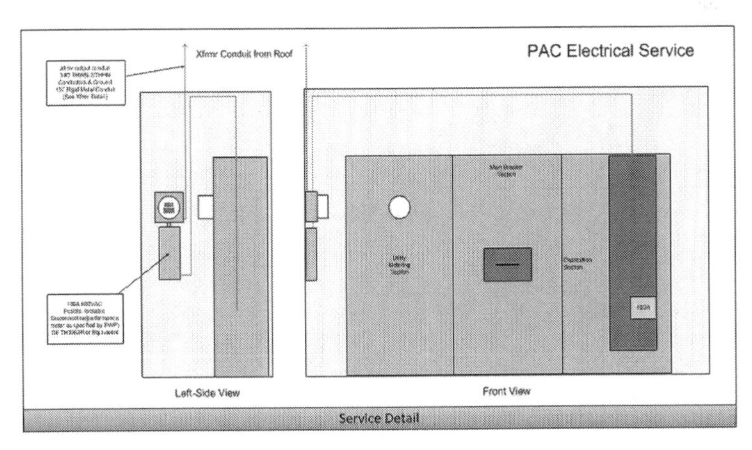

Figure 9 - Service Area Detail

Single-Line Drawing

The most significant deliverable in the rebate application packet was the Single-Line Drawing (SLD) for the electrical circuits. Since this drawing shows how all of the electrical components of the power generating system interconnect—including the interconnection to the utility's grid—we knew that this would be the most closely scrutinized piece of the submission. PWP has a generic SLD that installers can use (in fact, we helped develop it!), but that drawing does not cover the use of Enphase microinverters which we were featuring on this job, nor does it allow for a step-up transformer.

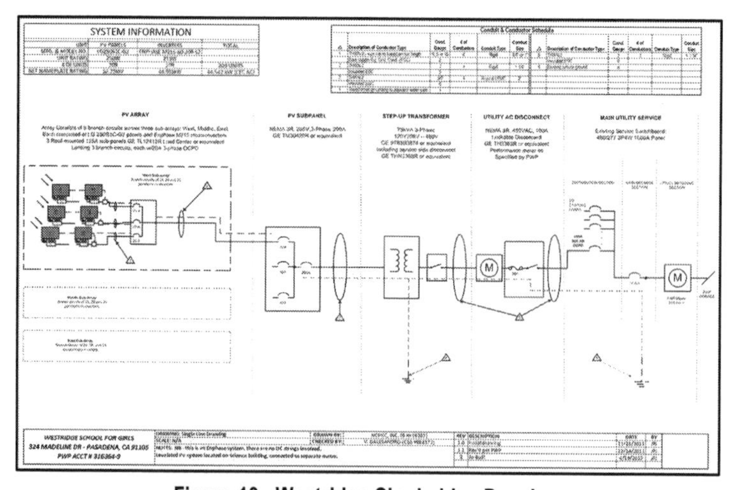

Figure 10 - Westridge Single-Line Drawing

Fortunately, we had developed a very flexible SLD format from prior jobs that we could readily adapt for this project. However, before we submitted it to PWP, we forwarded it to the application engineers at Enphase Energy to make sure that they were comfortable with what we had designed. Enphase was more than accommodating. Informed of our tight time frame, they bumped us to the front of their engineering review queue and came back promptly with the good news—

the design was good as we had drawn it, and no revisions were needed. Of course, that was no guarantee that the utility would agree, but it is always nice to have a Professional Engineer on your side!

Included in the SLD preparation was a complete set of voltage drop calculations. Voltage drop is a function of the resistance inherent in the conductors used (the larger the conductor the less resistance it has) and the amount of current you are trying to push through those conductors. As voltage drop increases, the system becomes less and less efficient. Indeed, if the voltage drop is too great, you can even have the microinverters shut off because the voltage that they "see" is outside of their operating limits. As a result, voltage drop calculations are necessary to insure that the system will have properly sized conductors to operate as intended.

Our calculations were complicated by the fact that we had 9 different branch circuits, three different sub-panels, and two different operating voltages! Good design calls for limiting total voltage drop to less than 2%. To keep the worst case scenario within that limitation (covering the branch circuit farthest from the main "solar-only" subpanel), we ended up with 4 different sizes of conductors at different legs of the run: #12 gauge in the branch circuit cables (as supplied by Enphase), #8 from the farthest branch circuit junction box to its corresponding rooftop-mounted subpanel, #2 from subpanel to ground-mounted main "solar-only" subpanel, #3/0 from that subpanel to the step-up transformer and then #2 again for the run from the transformer back to the service equipment area. (A client-requested design change during the install process increased the length of some of these runs—and that necessitated some wire size changes to insure that we stayed comfortably below the 2% limit.)

One Big Present

All of those documents, plus pages and pages of cut sheets describing all of the key products being used, were then submitted to PWP, one day before the deadline. With no margin for error, the submission had to be perfect. Thankfully, it was—PWP gave us their official blessing to proceed three weeks later, just three days before Christmas. One big present, indeed.

PULLING THE PERMITS

Whereas the rebate application is intended to demonstrate compliance with the local utility's requirements, the permit process is charged with guaranteeing that the proposed system, *as designed*, satisfies all applicable codes and standards. In theory, once you have successfully pulled the permit, the inspection process should simply be a matter of showing the inspector that you built the system according to the approved plans.

Figure 11 - This Roof Only Looked Conventional

This project presented one significant, but unanticipated challenge: the actual attachment of the system supports to the roof. Although the roof looked conventional enough, that was not a wooden truss underneath those shingles. To the contrary, the roof was built from a 20-gauge "Type B" steel deck with two layers of 5/8" plywood, followed by 3" of solid foam insulation, followed by 3/4″ of plywood to which the roofing materials themselves—membrane, felt and shingles—were attached. So the question arose: what would be a sufficient way to attach the standoffs to this building to provide the requisite resistance to the wind loads that would try to peel the array from the roof?

Wind loads are a function of several factors, including: the highest anticipated wind speed (called the Basic Wind Speed), the height of the building, the array's location on the roof, and the surrounding terrain. These are complex, and crucial, calculations in the design of a solar power system.

Figure 12 - Unirac FastFoot
© Courtesy of Unirac, Inc.

To handle that analysis properly, we turned to the structural engineer (SE) who had done the original load calculations for the building. Could we use a "FastFoot", and simply put multiple screws into the wooden decking materials? Surely with enough screws, and Unirac's FastFoot product, as Figure 12 shows, will allow for up to eight—we could reach the required pull-out resistance.

Unfortunately, that was not acceptable in this case because the SE could not personally vouch for the manner by which the plywood materials had been secured to the underlying steel deck. In other words, while we could be sure that with enough screws the array would remain attached to the plywood, we could not be sure that the plywood would remain attached to the building! Images of *Wizard of Oz* roofs, decked out in

50 kW of solar modules, flying through the air above Pasadena convinced us that we would need another way!

The SE suggested that we could use carriage bolts that ran all the way through the steel roof and were bolted together on the back side. Certainly such a scheme would guarantee that the array and the roofing materials stayed connected, and indeed, you would have to separate the steel deck from the steel framework of the building itself for that method to fail. Unfortunately, that approach was not practical, since there was no efficient way to access the back side of the roof in order to complete the connection.

Fortunately there was one other approach. A company called Triangle Fasteners sells some very strong, *very long*, self-tapping screws (called "Concealor screws") that could drill their way into the steel deck and provide us with the required pull-out resistance. The bad news was that all of the solar distributors that we could find only sold Concealor screws up to 7″ long, and that would not be long enough to guarantee that the screws made it through the decking.

A call to the manufacturer, however, revealed that in fact, they did make 8″ screws, they even made 9″ screws! We now had a solution that the SE could bless. It was time to pull the permits.

Fear and Loathing at the Permit Center

Anyone who has ever pulled a permit knows the combination of emotions that you encounter upon entering the building: fear that something you haven't considered will suddenly become ***A Really Big Deal***, loathing for the interminable waiting, and of course, the pain of paying for it all. Dentists' waiting rooms tend to be cheerier places.

Pasadena's permit center is certainly better than most: it is a comfortable old building across the street from the beautiful City Hall. They have a clever scheduling system that routes you among the different windows: Building and Safety, Zoning, Historical Preservation (very big in Pasadena but not a factor for solar projects), Fire, Permit Processing and, last but certainly not least, the Cashier. A solar project applicant must negotiate the required paperwork through every one of those windows before exiting with ***The Grail***: a stamped set of plans and a bright yellow permit folder where inspection sign-offs will be recorded.

First stop—Building and Safety.

Building and Safety

The building and safety department is responsible for reviewing the solar contractor's plans for conformity with state and local codes and standards—a really important task. Before you can make that happen, however, you have to speak with someone who knows what you are showing them, and on our first trip to the permit center, no such person could be found! The gentleman behind the B&S desk was very polite, and he seemed genuinely uncomfortable as he informed us that after our thirty minute wait, he couldn't help. Moreover, none of the people who "understood solar" were available—we would have to come back tomorrow.

Tomorrow dawned cloudy, but we were undeterred by the non-solar weather and determined to press forward. Our 35 minute wait this time was rewarded with an appearance before someone who was prepared to pass judgment on our plans! We walked him through each of the sixteen 24″ x 36″ pages, explaining as we went exactly what we were doing and where the answers to his questions could be found.

All seemed fine, but then he started throwing us some curves. The SE had done his calculations for a basic wind speed of 85 mph, the same wind speed we had always used for load calculations in Pasadena.

"No," he said, "You have to use ***100*** mph."

"Really? Since when?"

"Since the windstorm in Pasadena at the end of November," he replied. (Never mind that the wind speed never reached 85 mph in the city, let alone 100 mph, during that terrible event.)

"Really? When was that change adopted?"

"It wasn't," he conceded, but he nevertheless insisted that we needed to revise the calculations for 100 mph—or he wouldn't approve them. That meant another iteration with the SE and another trip back to the permit center.

The good news here was that we were certain that the attachment method would easily handle 100 mph winds (or 120 mph, for that matter), so this previously unannounced change in policy did not pose a danger to the project going forward. But changing the basic wind speed for an area from 85 to 100 mph is something of a big deal that will add to the expense of many future projects with no appreciable improvement in safety.

The other curve sent our way was really just odd.

We had done a detailed drawing showing the attachment method as it penetrated the various layers of roofing materials and made contact with the steel deck beneath. We drew that straight up on the page but also included multiple elevations in the sixteen page submission that showed the pitch of the roof and indicated that the array was installed on top of the attachment method, parallel to the roof.

"Not good enough," we were told.

"Why? What's missing?"

"You need to show the attachment at the slope of the roof."

"But we do. We show you the slope of the roof, we gave you the detail of how the attachment connects to the roof and we told you that the array is parallel to the roof. How is that not sufficient?"

"You need to add a drawing that shows the array attachment and which reflects the slope of the roof."

"Let me get this straight. What you want is for me to rotate the image of the attachment 13° to reflect how it will be pitched on the roof, and *then* you will approve this?"

"Yes," said the all-powerful man behind the counter.

Sigh. Ok, back to the drawing board (or more accurately, the computer screen). Fortunately, the SE was able to redo his calculations for a 100 mph wind speed in short order. And not surprisingly, it was also pretty easy to take the attachment image and rotate it. We shipped the revised plans to the printer and headed back to the permit center.

Surprise—there was yet another person behind the counter this time. Whereas his predecessor seemed to be actively looking for little things to complain about, this fellow could not have been more helpful. He looked at the revised load calculations—verifying that they had been done for 100 mph and that the SE had concluded that all was well—and then proceeded to stamp the plans. (I had pointed out the added, rotated drawing, but it was clear that he wasn't interested in that at all.) After he stamped the plans, ***he then took them himself to the zoning and historical preservation desks and secured those sign-offs as well!*** Wow! He saved us an hour of waiting in those queues, and he seemed genuinely concerned and eager to be helpful. What a pleasant contrast! We were well on our way with just one substantive hurdle remaining—the Fire department.

Fire

The California State Fire Marshall has developed a set of guidelines describing how fire departments should permit and inspect solar installations. The guidelines call for space to be set aside for pathways around the array and for venting of smoke in case of a fire. The guidelines call for different restrictions based on the size and shape of the roof and whether it is a residential or commercial building.

It is worth noting that while the document from the Fire Marshall was clearly labeled *"Guidelines"*, many localities seem to treat it as *Gospel*. Even more curious, the guidelines clearly say that they are just that—guidelines that do not have the force of law until a local jurisdiction passes an ordinance adopting the guidelines as ***regulations***. We have yet to see such an ordinance.

As originally designed, our building plan included a three-foot set aside around both sides of the array and from the ridge, and was augmented by automatic smoke ventilation devices already built into the roof. But that was deemed insufficient—the fire official insisted that we provide a *four*-foot clearance on all three sides. Yet another trip to the computer.

We returned with the revised drawings, showing four feet of clearance as requested. But now there was another complaint—the same fire official now wanted us to open a walkway in the middle of the array. (We already had access paths for potential maintenance, but they were not wide enough to be considered a walkway.) No matter that the roof was not at all like the flat roof with parapet shown in the guidelines, we still needed to provide a walkway. There was only one way to do that—take out a column of modules. Together we X-ed out seven modules and thereby created a walkway. The fire official was now satisfied, and she signed off on the plans.

Done

And just like that, we were done. Well, not quite—there was still the little matter of paying for all this. Here we made out surprisingly well. Unlike some cities that gouge solar permit applicants, Pasadena's fees were quite reasonable. Total permit cost for the now 52.3 kW solar project? $732. Sadly, we know of residential projects one tenth that size in other cities where the permit fees have exceeded $1,000! Fortunately, recent changes to California law should preclude such gouging in the future.[1]

Altogether, it took us four separate trips to the permit center, three plan revisions, and a little over $900 in expenses (including the cost of printing multiple plan sets and the permit itself) to secure our *Grail.*

Enough with the preliminaries, it was time to install some solar!

WORK ON THE GROUND

A project the size of what we were going to install at Westridge involves thousands of parts, all of which not only must be assembled properly for the system to work and be safe, but they must arrive in a timely fashion for the project to be completed on time. As we had access to the staging area for only two weeks while the school was on break, the project had a very narrow window in which to be completed.

By way of illustration, here's just a sample of the parts that were needed for this job: the roof attachments required hundreds of FastFoot plates, flashings, stand-offs, and flange connectors, (to say nothing of thousands of screws), on top of those attachments were mounted dozens of rail sections, end-clamps, mid-clamps, ground lugs and splices, and last, but not least, 209 micro-inverters and solar modules! Collectively these products came from five different distributors in four different states.

Needless to say, not everything goes as smoothly as you might like when you are pulling together all of these pieces. UPS likes to brag about Logistics, but we found some of their logistics to be highly *illogical.* Such as when they inexplicably sent two shipments that were sitting in an LA warehouse on a frolic and detour down to San Diego for the weekend, instead of driving them the twelve miles up the road to the job site.

Equally baffling were the folks who delivered the boom lift to the job site after-hours on a Friday evening without even a phone call and then just parked it out on the street—in front of a "No Parking" sign!

Eventually, everything arrived, whole and intact. As we prepared to begin the actual installation, the staging area was filled with:

Figure 13 - Staging Area

LG Modules, Enphase Microinverters, Unirac Rails & Lots of Wire

Transformation

The first task on the ground, once everything was on site, was to install the transformer which was required to step-down the voltage from the utility service (480 volts) to the voltage that would be used by our microinverters (208 volts).

Setting the Stage

The transformer was a 700-pound beast that had to be installed on a to-be-poured concrete pad in the equipment storage area on the East side of the building. To secure the transformer to the pad, we would imbed bolts into the pad and then maneuver the transformer on top of the bolts and anchor it with washers and nuts. Two key challenges had to be overcome: the first was to guarantee that the bolts were precisely positioned in the concrete pad since the transformer gave us very little margin for error.

Figure 14 - Transformer

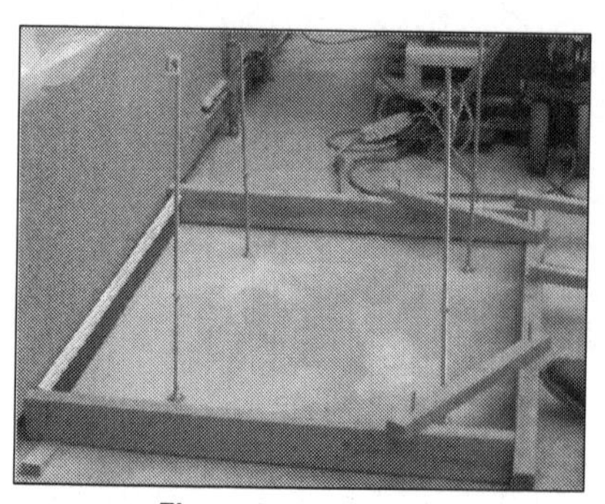

Figure 15 - Pad Prep

The second challenge was to get the transformer placed on top of the bolts without damaging them.

We solved the first problem by drilling into the existing concrete and securing our bolts into the ground with heavy duty anchors—as you see in Figure 15 with the framework for the pad surrounding them.

Then, when we were ready to fill in the form with concrete, we added some framing at the top to keep the bolts plumb.

Figure 16 - No Margin for Error

Once the pad was dry we maneuvered the transformer carefully into place, lowering a corner at a time until the narrow openings in the base of the transformer slipped over the (now cut to length) bolts. The final result provided a snug, but secure, installation.

The Art of Conduit

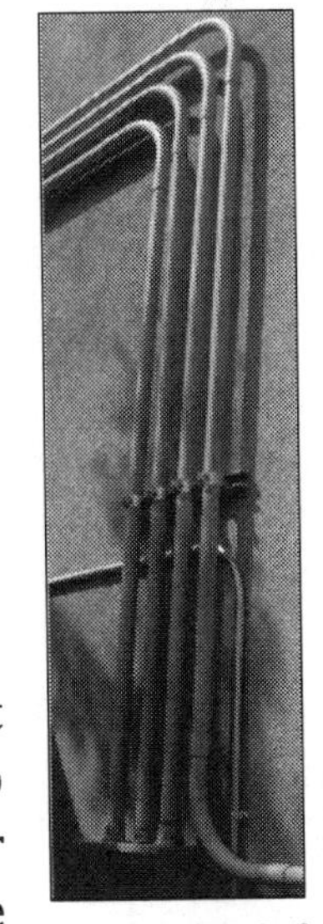

Figure 17 - Artful Conduit

In addition to the transformer, there were several other pieces of gear that had to be mounted on the ground including a 200 Amp subpanel, two disconnect switches, and a performance meter. Linking them all together was the conduit through which the conductors would be pulled.

Pasadena requires rigid metal conduit (instead of the thinner and easier to work with EMT) to be used for solar power systems wherever it is accessible on the outside of a building. That offers some additional safety, but it comes at a cost, especially given that we were using 1¼″ conduit for most of the runs. Rigid conduit of that dimension is heavy and cannot be bent by hand. Instead, a motorized pipe bender was the order of the day, and it took some really skilled craftsman to get the conduit in place and looking good.

It really is an art, as much as a science, and when done with care and precision, the result is quite appealing!

Pulling it Together

The final ground-based task was to pull the conductors through the conduits. The longest pull was 245′—not quite a football field, but close! Moreover, that longest pull had multiple bends as we routed the conduit to make it as invisible from the ground as possible. (To complete the task of making the conduit "disappear" to the greatest extent possible, the client painted the conduit two different colors to match the walls and trim!)

To facilitate the conductor pull we employed a powered wire "tugger" to assist us in coaxing the thick conductors through the long conduit run. Finally, at the end of a very long, drizzly Saturday, we were rewarded with having the conductors fully in place from the utility disconnect switch and performance meter socket to the disconnect switch adjacent to the transformer.[2]

The on-ground infrastructure well established, it was now time to commence work on the roof in earnest.

WORK ON THE ROOF

Solar projects actually begin on a computer screen as the designer tries to map what is known about the roof, the utility service, and the client's needs into a coherent proposal. As the project progresses through the rebate and permitting processes, that design is refined, and as we have seen, sometimes altered. But the trick of any implementation is to go from the designer's plan to an actual working system on the roof, and that starts with getting the attachments in place.

Figure 18 - Foreman & Designer Confer

Nearly ancient methods, consisting of tape measures and chalk lines, are the essential tools in this process. Since most solar arrays are essentially a grid, and this one was no exception, the trick is to project what is in the plans onto a corresponding grid on the roof. Precision and accuracy are the key to making this work, but roofs are notoriously inconsistent places. What seems to be square, isn't. What appears to be flat, actually is a series of peaks and valleys. Legitimate right angles and true parallel lines are scarce.

While the projection from design onto the roof proved easy enough, we were about to discover that what you see—or were told—isn't always what you get!

There's What You Plan and What You Get

As noted earlier, the biggest design challenge had been the need to account for the somewhat unusual roof construction at the site. We had been told that the underlying roof structure was a 20-gauge, type-B steel deck, overlaid with multiple layers of plywood, foam insulation and roofing materials. Given the thickness of those multiple roof layers we had determined that we would need to use four, 8-inch-long, self-tapping screws to secure the "FastFoot" anchors to the roof. We had purchased thousands of those specialty screws, along with a top-of-the-line Hilti-cordless driver, to do the job. We were confident that we were ready.

But then reality intervened.

As we started making the first few attachments we became concerned that not all of them were reaching the steel deck. ***In some places the actual thickness from the roof surface to the steel deck exceeded the 8" reach of the screws!*** A visual inspection from a scissors lift inside the building confirmed our fears: in some areas of the roof, only a few of the screws were penetrating the deck. On the other hand, in other areas all four screws penetrated the decking without difficulty. There was only one solution to the problem of the inconsistent roof, longer screws!

Fortunately, we were able to order some 9" screws from the manufacturer, the longest that they made. That did the trick—although the overnight freight charges needed to keep the project on track cost as much as the screws themselves. At least now we could be certain that every FastFoot plate was properly secured.

Lean on Me

The changes to our plans imposed during the permitting process meant that we were very tight on space. At the north edge of the array we had to install 3 subpanels, each of which had to handle three branch circuits that made up that sub-array. The original plan was to build a triangular cross brace out of unistrut to support the subpanels. Unfortunately, given the close quarters, the solar modules needed to come right up to the supports for the subpanels, meaning that a cross-brace system would take up too much space.

Instead, we designed a set of steel braces that were bent at precisely the angle that we needed—103°—to allow the subpanels to be perfectly vertical on the 13° sloped roof. The design was easy, but could we get them fabricated fast enough to keep the project on track? We knew of a small metal shop near our offices, and we took the design to them. Yes, they said, they could produce the six parts that we needed for $100, and they could have them ready for us in the morning—would that be soon enough?

Figure 19 - Roof-Mounted Subpanel

This turned out to be a very elegant solution to the problem. Using two FastFoot anchors, we attached unistrut to those anchors and then bolted the braces to them. When combined with the rigid conduit feeding the subpanel, we ended up with a very solid mounting solution.

Next problem!

Need a Lift?

The roof of the building was reachable by a series of three ladders traversing three different roof levels. While this was acceptable for getting personnel to and from the roof, it would never work for transporting hundreds of feet of rails, to say nothing of 209 solar modules!

Enter the boom lift — whether transporting rails, solar modules, or the Enphase microinverters as you see in Figure 20, the boom lift provided us with an efficient means of moving large amounts of gear up to the work site on the roof. Operating a device that articulates in multiple dimensions in relatively tight quarters takes skill and great attention to detail. (It also makes for some pretty cool looking photos!)

Figure 20 - Boom Lift in Action

Connect the Dots

Once the rails were installed, the Enphase microinverters could be mounted, and the process of running a continuous ground wire and the creation of the Enphase map could begin.

As with every Enphase system, we would be able to monitor the performance of the array down to the individual solar module/microinverter pair. (Indeed, this was a key selling point that distinguished our proposed system over the competition, because the monitoring ability meshed so nicely with the school's educational mission.)

Each microinverter had a serial number that was carefully peeled off and affixed to a "map" that showed where each inverter was located on the roof. As part of the commissioning process, we transferred the map data to the Enphase Enlighten website and built a true representation of how the system was laid out on the roof, down to showing the existing vents and designed in walkways within the array.

Now all we needed was to install the solar modules themselves. While this really is the easiest part of the entire process, it still involved many round trips on the boom lift, a very careful unloading and staging process, care to secure cables and connectors properly, torquing

Figure 21 - Solar Modules Going In

retaining bolts to their proper settings, and verifying that module/microinverter pairs were coming "online" as expected throughout the process.

Using the Enphase Envoy (the data logging and reporting device that uploads system performance data to the Enlighten website) and a laptop computer, we could verify that each and every module was properly connected and functioning as expected.

Thanks to that real-time data we could be confident that there would be no surprises that would need to be resolved later. Careful attention to detail during all of the previous, painstaking procedures that led up to this moment were rewarded during this final step with an array that aligned precisely and fit exactly as planned.

That gave us one last task for the boom lift—finished photographs. Here is one of our favorites:

Figure 22 - Westridge Project Complete

This project, commenced under a very demanding timeline, was completed with two days to spare, without Change Orders, and on budget.

Your project should be handled the same way.

GOING LIVE

Inspections

With construction complete, three inspections stood between us and officially going live with the system: by the city's fire and building & safety departments, and then by the local utility. When a solar contractor details precisely what they intend to do in the plans that they submit when pulling the permit, and then actually build what they said they would build, the inspection process should have minimal surprises. So it was here—the inspection process provided no drama.

We had agreed to eliminate a column of modules as part of the permit process at the fire department's request, and, not surprisingly, that was the primary thing they wanted to see. Once we showed them that the "as-built" system matched what had been approved in the permit office, the sign-off was immediate.

Similarly, the inspections by the Building and Safety department and PWP went forward without a hitch. Once we had secured all three sign-offs, the PWP performance meter was installed, and we were given permission to bring the system online.

One of the highlights of this particular system is the monitoring that is provided by the microinverter manufacturer, Enphase Energy. Figure 23 shows how the system appeared on the Enphase Enlighten website one recent sunny day:

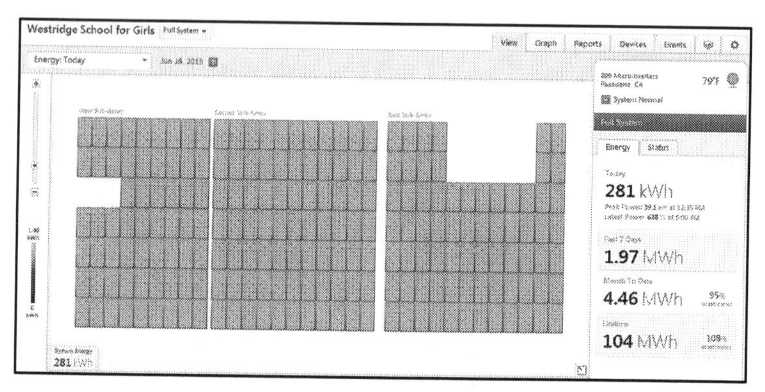

Figure 23 - Westridge System Visualized by Enlighten

The students at Westridge will be able to analyze the performance of this system for years to come,[3] and it will provide them with a first-hand experience of how renewable energy works and how it can make a difference in our lives—what great lessons to learn!

Rebate Processing

The final step was to submit the Incentive Claim Form to allow the client to begin receiving annual, PBI-rebate payments. The bad news? PBI rebates are only paid after the system has actually produced power over a period of time. In PWP territory, that means no payment at all until after the end of the first year—a long time to wait. The good news? The Westridge system out-performed our original estimates, and as a result, the rebate they received was higher than predicted.[4]

Now that is the kind of surprise that would warm the heart of any facilities manager—and his CFO!

Chapter Notes

[1] California's law limiting solar permit fees caps most residential projects at $500 and provides a formula for commercial projects based on system size. The bill, SB 1222, was signed into law by Governor Jerry Brown on September 27, 2012. http://leginfo.legislature.ca.gov/faces/billNavClient.xhtml?bill_id=201120120SB1222, accessed 6/16/2013.

[2] Interested readers can see more photographs from this installation by visiting the Run on Sun Facebook page albums where even more details from the Westridge project are shown. https://www.facebook.com/RunOnSun/photos_albums, accessed, 6/23/2013.

[3] One novel use for the data coming from the Westridge Enphase Enlighten website was to analyze the effect of a partial solar eclipse. See, *Solar Eclipse – A Data Geek's View,* http://runonsun.com/~runons5/blogs/blog1.php/solworks/solar-eclipse-data-geek-view, accessed 6/23/2013.

[4] See the discussion of the Westridge rebate in Second Case Study: Westridge School for Girls, Pasadena, California – The Beauty of Doing it Right, p.117.

VI. SPECIAL CASES

Jack's experience was fairly common for a small, commercial solar power system that is roof-mounted and interconnected to the grid by way of a net-metering agreement. But solar is not a one-size fits all solution; there are a number of special cases that are starting to appear that provide other ways to add commercial solar.

Four of these special cases will be discussed here: Solar Carports and Other Shade Structures (that is, specialty structures that are created expressly for the purpose of hosting a solar power system), Feed-in Tariffs (whereby a solar power system owner is paid by the utility for all of the energy produced by the system), Community Solar (which allows a locally-sited solar asset to be shared among a number of utility customers), and the growing role for Local Energy Storage (which has the potential to take the bite out of demand charges and help commercial solar become a leading energy source going forward).

CARPORTS AND OTHER SHADE STRUCTURES

While most commercial solar power systems are located on a building roof, there can be a host of reasons why that is not desirable. For example, the roof might be obstructed by pre-existing equipment (commonly HVAC units) that would interfere with the array's layout, or the roofing material might be nearing the end of its useful life but the building owner does not want to re-roof prematurely, or there could even be structural concerns regarding the roof's ability to support a commercial solar system.

When any of those issues arise, a potential alternative means for installing solar is the free-standing carport or shade structure to which the solar array is attached.

Figure 24 - Solar Carport by Baja Construction

These structures are becoming more and more popular as they convert an otherwise unproductive asset—a series of parking spaces—into a revenue stream that also serves as a declaration of a company's intent to go green while improving its bottom line. Plus, they keep the cars parked underneath cooler during the summer, a side-benefit greatly appreciated by those who park there.

Carport structures typically extend over one or two rows of cars per structure and can extend in width from just a few parking space to dozens. Typically a structure designed to shade a single line of cars can support five to six rows of solar modules, meaning that the total array size can grow very rapidly.

More general shade structures are not confined to the linear arrangement of a carport and are really only limited by the site's footprint, the designer's imagination, and the project's budget.

Of course, adding a specialty structure to support a solar array adds cost and complexity to the project. As a general rule a specialty sub-contractor—like Baja Construction, the company that built the solar carport shown in Figure 24—will be brought into the project to handle the actual construction, and the design will have to go through its own permitting process. Depending on its location, such a system might incur significantly greater costs for conductor sizing (to avoid excessive voltage drop), conduit runs and trenching—particularly if an existing concrete parking lot needs to be saw cut to allow for the trenching. Add to that the cost of the structure itself and the added costs can become significant quickly.

Nevertheless, as the cost of solar continues to decrease, and more carport contractors enter the market, thereby driving down the cost for the structures themselves, we anticipate seeing more and more acres of parking lots transformed into solar arrays.

FEED-IN TARIFFS

A Feed-in Tariff (FiT) is a program by which a utility pays the owner of a solar power system for every kilowatt-hour of energy the system produces. It was the introduction of an extremely generous FiT program in Germany that resulted in that sun-starved country becoming the world leader in installed solar capacity.[1]

A FiT program contrasts with a net-metering agreement in that the owner of a FiT-supported system does not consume any of the energy produced locally (*i.e.*, to power the site's own loads) and then provide the excess to the utility. Rather, under a FiT the solar power system feeds ***all*** of the energy produced directly onto the utility grid, and the system owner does not offset any of their existing usage. As a result, there is no reduction in the site-specific utility bill; instead, the utility sends a check for the value of the energy produced to the system owner each month. Nor does a FiT-supported system receive a rebate from the utility (although they still qualify for tax incentives assuming they are a for-profit entity).

A FiT-supported solar system can be economically attractive where conventional net-metering is not, such as at large sites with relatively low usage or sites that have multiple tenants who each pay for their own electric usage via a separate meter. In such cases the property owner can convert unproductive space—the building roof or possibly carports in the parking area — into a valuable asset that produces a long-term revenue stream.

The owner of a solar power system subject to a FiT will contract with the utility for an extended period of time (typically 20-25 years). During that period, the utility will pay the system owner a Base Price for Energy (BPE) which may, or may not, be subject to Time-of-Delivery (ToD) multipliers. To determine what the utility will pay for any given kilowatt hour produced, you must multiply the BPE times the applicable ToD multiplier. Thus, to predict accurately the revenue stream associated with a FiT-supported solar system, the developer will have to model the system's output on an hour-by-hour basis over the course of an entire year.

Design Matters – A Survey of Some Southern California FiT Programs

A FiT program, by definition, is supposed to be open to all qualified applicants on a first-come, first-served basis. In Southern California, the FiT programs that have been developed all have total Megawatt capacity caps associated with them, so a potential applicant runs the risk of the only available FiT program being fully subscribed before they can apply. Since such programs are tied to the local utility, the applicant is not allowed to sell their energy to some other willing utility buyer.

Beyond the matter of timeliness, the overall design of the program matters a great deal. Although the most important question is how much is going to be paid for delivered energy, other factors come into play in assessing the desirability of an offered FiT. Among those factors are:

- Predictability—how accurately can the developer know what they will earn based on when they decide to join the program;
- Transparency—are all of the costs of participating known in advance, and, if not, are

there appropriate off-ramps built into the process to allow the developer to exit gracefully if newly discovered costs (such as for interconnection) undermine the project's profitability; and

- Reliability—will the utility support the program in the manner that they have indicated that they will, and what recourse does the developer have if they do not?

As one utility official noted, "it is one thing to develop a program that complies with the letter of the law, it is another thing to develop a program that works." [2]

LADWP's FiT Program

The Board of the Los Angeles Department of Water and Power (LADWP) approved their FiT program on January 11, 2013, and the LA City Council followed suit ten days later. The initial program calls for the allocation of 100 MW of solar capacity over the next two-and-a-half years. Project sizes can range from as small as 30 kW to as large as 3 MW.

Every six months, starting February 1, 2013, LADWP will open the next tranche of 20 MW of solar capacity. The initial BPE was set at 17¢/kWh and will step down one cent in each subsequent tranche. If an entire tranche is not fully subscribed during the six month window, that capacity will fall out of the program (*i.e.*, it will not be rolled into the subsequent tranche), and the next tranche will be opened on schedule at the lower BPE. In each 20 MW tranche, 4 MW are set aside for "smaller" projects—specifically those between 30 and 150 kW. (These smaller projects also qualify for lower application fees.) If the entire 4 MW is allocated before the 6 month window closes, smaller projects can tap into the remaining 16 MW (assuming that hasn't been fully allocated to larger projects.) Thus, the program provides an application carve-out for smaller projects, helping to improve the likelihood that such projects are proposed and approved.

To prevent a stampede at the application window, LADWP determined that all applications received during the first five business days of a new tranche will be ordered by way of a lottery. Applications received after the first five business days will be date stamped and processed in the order received.

The first tranche, offered in February 2013, sold out immediately for "large" projects and the "small" projects category was fully subscribed within two weeks.[3] The second tranche, offered in July 2013, was subscribed just as quickly.[4]

Based on that performance, it is likely that the next couple of tranches will also sell out almost immediately, given the relatively high prices being paid and the pent-up demand for installing non-residential solar power systems in LADWP territory.[5]

LADWP's FiT program incorporates a six-component Time-of-Delivery (ToD) multiplier table that alters the amount paid for delivered energy from 225% of BPE (during High Peak in High Season) down to as little as 50% of BPE (during the "Base" period).

Our analysis indicates that these ToD multipliers can increase the annual compensation paid to a site owner over just the BPE from 5% to 10% depending on how well the system design takes advantage of the site characteristics.

As complicated as it is, LADWP's FiT program reflects the end result of years of planning and a multitude of public meetings to solicit input from all of the affected stakeholders. The initial success of the program—over-subscribed in both the large and small project categories—is a testament to a program that was "designed to work."

FiT Programs in Anaheim and Riverside

The cities of Anaheim and Riverside also have FiT programs, but their experience has been vastly different from what has taken place in LADWP territory.

Both programs "feature" a BPE, but unlike the table published by LADWP that clearly lays out what will be paid for the entire life of the program, the Anaheim and Riverside programs allow for the price to change in a potentially unpredictable manner each year.

The BPE in Anaheim is a paltry 4.6¢/kWh whereas in Riverside it is 5.8¢/kWh—neither of which is intended to encourage developers to build projects, and that is precisely what has happened. In the two years that the programs have been in effect, ***not a single FiT project application has been submitted!*** In other words, both programs were effectively designed to fail.

Glendale Follows Suit

The last local municipal utility to comply with the state law mandating FiT programs by July 1, 2013 is Glendale Water and Power. Having had the benefit of watching all three of these programs unfold before it, you might expect that Glendale would pick and choose the best elements from each, thereby coming up with a program that combined the predictability and transparency of the LADWP program with the lower fees required by Anaheim and Riverside.

But you would be wrong.

In what can only be described as a "worst of all possible worlds" approach, Glendale has concocted a program that includes a fee structure as high as it is in Los Angeles, but offers miserly energy prices ranging from 7.3¢ to 9.3¢/kWh (depending on time of delivery). Worse still, that price can change every quarter, solely at the discretion of Glendale! In other words, a potential project developer is being asked to sign a long-term output contract where the only permissible purchaser of their energy is Glendale's utility, but the utility, at its sole discretion, can revise the price paid for that output. Who would enter into such a contract?[6]

FiT programs have great potential to incentivize the wide-scale adoption of solar, but as long as the design of those programs is left to the whims of the utilities, finding a program that was "designed to work" apparently will remain the exception and not the rule.

COMMUNITY SOLAR

While a commercial solar power system is a great fit for a company like EnGex, many businesses occupy leased space with property owners that are unwilling to allow solar to be added to their property. Historically that would have left such a company with no access to solar, however, there is a new program coming that has the potential to meet the needs of such companies, and many others.

The program, known as Community Solar, allows a solar developer to create a solar project and then sell pieces of its output to a host of subscribers. Under such a program, a solar project is built and individual utility customers—homeowners, renters, business owners—each purchase a portion of the system's output. The solar developer sells the power directly to the utility and the individual participants receive a credit on their bill in proportion to their share of the system's production.

For example, let's say a typical Pasadena homeowner needs a 5kW system to offset their usage whereas a renter needs 2.5kW and a local small business needs 20kW. A solar developer builds a 200kW system (maybe on a warehouse roof or a vacant lot), and sells corresponding shares to 8 homeowners, 16 renters, and 6 small businesses. Everyone gets precisely what they need with no concerns about local shading, or roof ownership, or even the possibility of a leaking roof!

In addition to the benefit for commercial operations that cannot install solar on their businesses, it is estimated that as many as 75% of all households cannot presently add solar due to ownership or space limitations,[7] meaning that Community Solar could greatly increase the number of utility customers who could take advantage of solar, thus driving down the cost of solar for everyone.

Unfortunately, the utilities are opposed to a threat to their business model that not only undercuts their ability to sell their product, but also puts them in the position of having to account for the product sales of a competing third-party. As a result, Community Solar legislation was defeated in California in 2012,[8] and the fate of revised legislation, pending as this is being written, is unclear.

THE RENEWED ROLE FOR LOCAL ENERGY STORAGE

Back in the days when most solar systems were designed to operate "off-grid" (and long before the introduction of net metering), energy storage in the form of large banks of batteries was a standard component of solar power systems. Since very few commercial operations take place off-grid, net-metering policies replaced battery backup systems as the essential component to making commercial solar economically viable.

Yet even with net metering, there are some problems for conventional commercial solar power systems. For one thing, many commercial customers operate under utility rate structures (such as SCE's GS-2 or PWP's M-1) that include substantial demand charge components. A site's peak demand occurs at a time of day that will vary from site to site based on the specifics of how it is utilized. For example, a commercial office building may see its peak demand during the afternoon when HVAC systems are struggling to cool the building. On the other hand, a Church building might see its peak demand on Sunday mornings (when all the lights and HVAC are turned on at once), whereas a water company's pumping station might experience its peak demand in the middle of the night.

If the timing of the peak output from the proposed solar power system is out of sync with the site's peak power demand, and the site owner cannot switch to a more solar-friendly rate structure, much of the economic advantage of adding solar may be lost.[9]

In addition to concerns regarding demand charges, another challenge to commercial solar arises from utility pushback in the form of an attack on net metering itself as being fundamentally "unfair" to utility customers who cannot afford to install solar themselves.[10] Leaving aside the questionable credibility of investor-owned utilities acting as advocates for their poorer customers, the populist rhetoric coming from these utilities will certainly find an audience in certain policy circles. The resulting increase in pressure on regulatory bodies to revisit net-metering programs poses an existential threat to those programs. But *any* reduction in the value of the credit derived from net metering will also erode the economic appeal of commercial solar.

Beyond those economic concerns, even progressive solar policy states like California have adopted caps on the amount of net metering that is allowed,[11] arguably due to concerns about the impact of high amounts of solar penetration on the stability of the local grid. Since solar cannot be controlled by the grid operator—that is, it is not *dispatchable*—large amounts of distributed solar generation has the potential, the argument goes, to create grid instability and possibly even trigger blackouts.

Intelligently controlled, affordable local storage is a solution to all of these problems, and it is unquestionably the hottest topic in the solar industry today.[12]

Unlike the conventional lead-acid batteries and relatively simple charge controllers found in off-grid or hybrid systems in the past, this new generation of local storage technology combines state-of-the-art energy storage systems with sophisticated load management to eliminate demand peaks, shift energy availability from production to need, and even provide for curtailment—the ability to limit or eliminate system output remotely—if necessary for grid stability.

The concept is simplicity itself, but the devil is in the details, and in the price. Excess energy from the solar power system is stored in a local repository: lithium-ion, flow or other battery technologies. As production falls below need, excess energy is drawn from the storage device instead of the grid. If a power demand spike occurs outside of peak solar production, that spike is satisfied—in whole or in part—by drawing on stored energy, thereby reducing the peak that drives demand charges.

While these solutions have been hinted at during solar trade shows for several years, off-the-shelf systems that are cost-effective from known manufacturers are not yet available. However, given the clear need for such systems and the influx of investor dollars into their development,[13] we anticipate that local storage options will be not only available, but common, in the next five years.

Chapter Notes

[1] CLEAN BREAK: THE STORY OF GERMANY'S ENERGY TRANSFORMATION AND WHAT AMERICANS CAN LEARN FROM IT, by Osha Gray Davidson, InsideClimate News, 2012.

[2] Quote from Ms. Carrie Thompson, an Integrated Resource Planner for Anaheim Water and Power and their FiT coordinator, personal communication, June 20, 2013. Under California law (SB 1332), publicly owned utilities (also known as municipal utilities) with more than 75,000 customers were required to offer a FiT as of July 1, 2013. http://www.leginfo.ca.gov/pub/11-12/bill/sen/sb_1301-1350/sb_1332_bill_20120927_chaptered.html, accessed, 6/23/2013.

[3] See, *LA Feed-in Tariff Update: First Tranche is Fully Subscribed*, by Jim Jenal,

Chapter Notes, *continued*

http://runonsun.com/~runons5/blogs/blog1.php/solecon/feed-in-tariff/fit-update-first-tranche-subscribed, accessed, 2/16/2013.

[4] See, *LADWP FiT Second Tranche Report,* by Jim Jenal, http://runonsun.com/~runons5/blogs/blog1.php/solecon/feed-in-tariff/ladwp-fit-2nd-tranche-report, accessed, 9/3/2013.

[5] LADWP considers its FiT program sufficiently successful that it has phased out solar rebates for non-residential customers, leaving non-profit entities—for whom the FiT doesn't pencil out given the lack of tax incentives—at a decided disadvantage. See, *LA Non-Profits Bid Solar Goodbye*, by Jim Jenal, http://runonsun.com/~runons5/blogs/blog1.php/non-profit-solar/la-non-profits-bid-solar, accessed, 6/7/2013.

[6] Apparently no one is willing to sign-up for Glendale's FiT. As of September 3, 2013, the application queue at the Glendale FiT website was empty, http://www.glendalewaterandpower.com/pdf/Feed-In-Tariff_Queue_2013-2014.pdf, accessed 9/3/2013.

[7] 75% of households cannot take advantage of solar presently, from Bill Summary for SB 43 & AB 1014, http://californiasharedrenewables.org/wp-content/uploads/2013/03/SB-43-and-AB-1014-Bill-Summary.pdf, accessed, 6/23/2013.

[8] *Community Solar Bill Defeated – Wait 'til Next Year*, by Jim Jenal, http://runonsun.com/~runons5/blogs/blog1.php/ranting/community-solar-bill-defeated, accessed, 6/23/2013.

[9] See the discussion of demand charges and utility rate structures in Chapter II. Preliminaries: What You Need to Know First: The Basics—Usage & Demand, p. 10.

[10] *It's Time to Reevaluate Net Energy Metering for Solar Power*, by Helen Burt, Chief Customer Officer, PG&E, http://www.pgecurrents.com/2013/02/05/helen-burt-

Chapter Notes, *continued*

it%E2%80%99s-time-to-reevaluate-net-energy-metering-for-solar-power/, accessed, 6/23/2013.

[11] California's cap on net-metering agreements is 5% of a utility's "aggregate customer peak demand." See, *Decision Regarding Calculation of the Net Energy Metering Cap*, California Public Utilities Commission, issued 5/30/2012, http://docs.cpuc.ca.gov/word_pdf/FINAL_DECISION/167591.pdf, accessed, 6/23/2013.

[12] See, *Intersolar 2013 – The Storage Debate: Poseurs and a Player*, by Jim Jenal, http://runonsun.com/~runons5/blogs/blog1.php/solworks/est/intersolar-2013-the-storage-debate, accessed, 7/20/2013.

[13] Bill Gates Gives Boost to Renewables Storage, by Ben Willis, PVTech, http://www.pv-tech.org/news/bill_gates_gives_boost_to_renewables_storage, accessed, 6/23/2013.

VII. LESSONS LEARNED: REPORTING FROM THE ROOFTOPS

Just as Jack tried to tap into sources of real-world knowledge, we would be remiss to conclude this step-by-step exploration of adding commercial solar without looking at a couple of case studies that show how this can work when done right, and the risks inherent in doing it wrong.

FIRST CASE STUDY: ENVIRONMENTAL LABORATORY, ALTADENA, CALIFORNIA – THE HAZARDS OF DOING IT WRONG

More and more frequently we are asked to come out and evaluate existing solar power systems that were built by others—sometimes because there is a clear problem and the original installer is no longer in business, and sometimes because the system owner simply wants to get an independent assessment of the performance of their system.

This first Case Study arose from one of those latter instances. The system owner was puzzled by how high their post-solar bills were. Surely if the system were producing all of the energy that they had been led to expect, they should have been seeing much lower bills. (Frankly, we expect to see many more of these cases in which systems have been oversold or bills have been under-explained. Installing solar is, at least in part, a process of managing expectations with candor, but too many installation companies don't seem to know that.)

When we got to the site, what we saw *seemed* very impressive indeed.

Figure 25 - An Impressive Looking Carport Installation

Eight carport structures were arrayed around the building, facing south and west and tying into a nicely assembled collection of SMA inverters. Conduits came up from the ground into a neatly mounted gutter to feed each of the eight inverters. This was a well-designed piece of work—and everything appeared to be operating properly.

There were two more sub-arrays on the roof of the building, and there too, everything initially appeared to be in order.

Figure 26 - Roof Arrays

Remember, we were only there to take measurements and assess the performance of the system—and from a system performance perspective, everything seemed fine.

But then this bit of "engineering" caught my eye:

Figure 27 - What is This?

What was going on here?

This was a system built less than a year ago, and yet they had chosen to construct their tilt array using *ad hoc* pieces of bolted-together unistrut? That certainly didn't seem up to the construction standards of the carports, and it wasn't something that we would have done, but presumably an inspector had climbed up the same ladder I had and blessed this as structurally sound—right?

Feeling more than a bit concerned, I decided to go around to each rooftop-row and make sure that everything was still secure.

That's when things really went south. Nearly a third of the rows had one or more module corners that were no longer secured to the unistrut—in some cases both north and south edges were free, meaning that the only thing holding that module to the "rails" was a pair of mid-clamps that, under stress, could not possibly prevent sideways motion that would allow the modules to come completely free! Given what we had seen a Pasadena windstorm do to an array that was far better secured, it was very disturbing to imagine the future fate of this nearly-new array.

Upon closer inspection we figured out how the contractor who had built this system *intended* it to work.

Figure 28 - Secured by a Wish and a Prayer

At the bottom of Figure 28 is the piece of unistrut serving as a mounting rail. On the left we can see the lip of the solar module and holding it against the rail is a bolt that is screwed into a channel nut in the unistrut. But that bolt doesn't pass through the module frame (even though the manufacturer has provided pre-drilled holes precisely for that purpose). Instead, the only thing keeping this module in place is the friction between the lip and a fender washer which is entirely unsupported on the right side.

This is laziness or ignorance of staggeringly dangerous proportions. How hard would it have been to adjust the unistrut to align with the module's pre-drilled holes? But the wizards who built this array either didn't realize that would be a good thing to do, or they simply couldn't be bothered to do it!

Figure 29 - Unsecured

The ultimate results of this shoddy work were everywhere to be seen. Here the washer has slipped away from the lip of the module and now that edge of the module can easily be lifted from the unistrut rail—by hand, or by the wind.

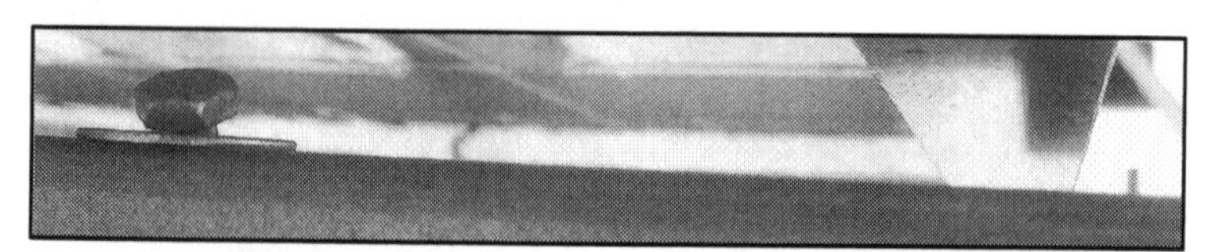

Figure 30 - Miss Three Inches or a Mile - Same Result

Finally, in Figure 30 the bolt and washer is far to the left of the module lip that it is supposed to be securing to the unistrut. Did the wind over the past year move it that far away? Or was it never even put in its "proper" place by the install team? Either way, there is nothing holding this module to the unistrut on this end.

Keep in mind—as you can see from our first picture of the roof array—these modules are exposed to the full force of the wind coming out of the North. In less than a year of service all of these failures to secure these modules have developed. What will happen when the next big windstorm sweeps into town?

Pity the company that owns this system. They paid top-dollar to a supposedly reputable contractor to design and install this system, and now they have a crisis on their hands through no fault of their own. Indeed, once down on the ground we took a closer look at the carport structures, and realized that the same mounting "technique" had been used to secure the modules to the steel members of the carports! In other words, ***all of the edge-mounted modules were in danger of coming off of their respective carport structures!***

What lessons can we learn from this badly botched installation? Some failures occur because of mistakes or even bad luck—but such was not the case here. The crew that built this system either utterly failed to comprehend the nature of their task, or they simply didn't give a damn. (And again, this site was inspected, so what was the inspector looking at before signing off on this?)

The company that paid for this installation is a highly reputable and professional organization, and they ob believed that they were hiring an equally professional installer—but they certainly didn't get what they paid for.

How might they have protected themselves? One possible hedge would have been to hire a third-party consultant—possibly a NABCEP certified installer or a professional engineer—to look over the installation before making final payment. That at least would have given them some notice of these problems before modules were ready to start flying off the roof.

Indeed, perhaps the greatest irony of this entire episode was that the system was actually performing extremely well, at least for now. The reason that we had been called out to the site in the first place was not an issue at all (except in the solar company's failure to predict accurately the output of the system that they installed). But then, all too often, when things fail, they do so in ways that we never imagined.

SECOND CASE STUDY: WESTRIDGE SCHOOL FOR GIRLS, PASADENA, CALIFORNIA – THE BEAUTY OF DOING IT RIGHT

For the past 100 years, Pasadena's Westridge School for Girls has prided itself on providing a top-notch education that prepares today's young women to be tomorrow's leaders. As part of a legacy of bringing cutting-edge ideas and innovation to campus and making it all come alive in the classrooms, school officials have made it a goal to pursue and complete at least one project related to sustainability on campus each year.

In late 2011, school officials decided to address a perennial request made by parents and students over the years: that an air-conditioning system be installed in the school gymnasium to keep crowds cool on hot days. While the school could likely raise funds from the parent community to purchase and install a system, the financial and environmental impact of running it and creating a new drain on the school's power supply were less than sustainable, according to Brian Williams, Westridge's Director of Facilities.

To offset the cost and energy consumption of the desired air-conditioning system, officials decided to fundraise for another project at the same time, asking parents and students if they would be willing to raise money to install a commercial-scale solar power system on school grounds.

Big Plan on Campus

Solar energy was not exactly new to the school. In 2009, Westridge had installed a small system on its newly built LEED-Certified Platinum science and math building. There some of the panels were placed near the ground, where students could see them up close and interact with the technology.

For the new project, however, school officials were thinking much bigger. They wanted to maximize their energy savings by installing a larger system that would provide a solid and swift return on their investment and take advantage of a local rebate being offered by Pasadena Water and Power (PWP) that was set to decrease drastically by the end of the year.

Ideally, the new installation would generate enough energy to cover the usage from the HVAC in the gym and still supply additional power for other buildings on the campus, saving on the school's overall monthly electric bill. At the outset, there were no specific parameters for the project, Williams said, just a location.

"We found our biggest roof and said, 'Fill it up.'"

The south-facing slope of the roof of the Fran Norris Scoble Performing Arts Center would be the perfect place to set up a large installation. With an area of approximately 4,000 square feet, it received direct sunlight throughout most of the day, and was elevated enough to keep the operation tidily out of sight.

Doing their bidding

Williams turned to three different solar companies, seeking estimates to help define the scope of the project and determine a total cost that would guide the school's fundraising efforts.

The first two companies Williams considered were those the school had worked with on past projects. One was a large, international solar provider that had worked on the 2009 science building installation. The other had previously worked with Westridge on nearby residential properties.

Run on Sun, the third provider approached regarding the installation, was the only one the school had not personally worked with before. The Pasadena-based company had contacted school officials earlier that year and delivered a presentation on its installation process, qualifications and services provided. That presentation was fairly comprehensive, and made enough of an impact for Westridge to include Run on Sun in the bidding process.

"We didn't have any actual experience with them, but they had very good references that we checked as we went through the bidding process," Williams explained.

Finding the right contractor for the job was a thorough search that required analysis and research. But beyond that, Williams—who's worked in the facilities department for more than two decades—relied on his own gut instinct and experience to determine which company could best get the project done in a short amount of time and leave no detail unchecked.

Since Westridge is a privately funded nonprofit, it is not under the same obligation as public entities to accept the lowest contractor bid. Cost, however, was still a primary consideration in selecting an installer.

"I have a fiduciary responsibility in my job to make sure the projects we do here are cost-effective and help maintain our campus in a positive way," Williams says. "I can't do anything that doesn't make fiscal sense."

The cost estimates provided by the three companies for an approximately 50 kW system were not that disparate from one another, coming in at about $4.25 per watt. One thing Westridge did have to be cautious about in its review of the estimates, however, was the possibility of a provider initially underbidding the project with the intention of requesting additional funds through change orders after a contract had been signed. The contractor who was truly right for the job would be one who provided an accurate estimate up front and diligently held true to that figure.

When considering the PWP rebate, Williams knew the school stood to save about half the total cost of the installation over the course of five years. Additionally, all three bidders told Westridge officials they could expect to see a full return on their investment in about seven years. Overall, the project made great financial sense.

Since the cost estimates that came in from the three bidders were relatively similar, Williams also analyzed each company's references, experience and credentials. For example, he considered installers certified by the North American Board of Certified Energy Practitioners (NABCEP) to have an advantage over uncertified companies.

Throughout the selection process, Westridge officials knew they were working under a bit of a time crunch. If they missed the window to apply for the PWP rebate, the project would end up costing significantly more than the amount they had raised. Because of this, hiring an installer who could successfully submit all the paperwork before the fast-approaching deadline, and one they could trust to get the application process 100 percent right the first time, was of critical importance.

In this regard, Run on Sun was a standout, Williams says. They promised to take care of all the paperwork, filing within the deadline period and eliminating the extra bureaucratic steps the school would otherwise have had to take.

"They managed the whole process—I basically [just] had to sign the form," he says.

The fact that Run on Sun had a strong working relationship with the city of Pasadena was another important consideration that worked in its favor, given Westridge's own solid standing in the community. The attention to detail and professional accountability demonstrated by the company throughout the bidding process brought the local provider to the top of the list, making the ultimate decision to go with Run on Sun an easy one, Williams says.

"It was a combination of the cost, return on investment, their relationship with the city, the impression we had with the provider, how they presented the information, and their references. We looked at all of it."

Run on Sun Gets it Done

With the paperwork filed and the rebate secured, Run on Sun had a very tight window— less than two weeks in April 2012—in which to install and connect all 209 solar modules and get the monitoring software up and running. The goal was for the project to be completed and operational by the time Westridge students returned from Spring break, so time was of the essence.

For the rooftop array Run on Sun used a micro-inverter system supplied by Enphase Energy which allows customers and solar installers to track the output of the modules, individually or collectively, from the convenience of their computers, iPads or smartphones. This software was a selling point for the school, because it would make the technology accessible to students and allow teachers to incorporate aspects of the solar system's performance into creative classroom instruction.

The modules were grouped into three sub-arrays that formed a larger circuit. Under each module, a microinverter was installed to convert DC power gathered from the module into AC power, which could be combined and fed back into the school's electrical service. Having a 1:1 ratio of microinverters to modules allows for a more detailed readout that lets users know the output of each module, and gives an easy-to-read display should anything ever go amiss, from a connection issue to dirt on the module's surface.

Throughout the installation Williams remained on hand to oversee the work, though there were no delays and no change orders requesting funds beyond what had been originally estimated. Within the assigned two-week period, Run on Sun had completed the project on time, and everything was in working order.

"I've worked with a lot of contractors, and I can honestly say, in this situation, this was one of the most seamless projects we've ever completed," Williams recalls. "They were here early on the first day and, boom, they got it. It was done on schedule, at the price they said and signed off by the city. I wouldn't hesitate to do a project like that again."

A Truly Happy Anniversary

April 2013 marked the one-year anniversary of Westridge School's solar installation, and Williams reports the system is running smoothly. The Enphase software makes it easy for officials, teachers and students to monitor the activity of all 209 modules, but Run on Sun also keeps a close eye on the operations via remote monitoring, and reaches out if and when an anomaly is detected. In the event that an outage or a decline in energy production should occur, the company promptly notifies the school.

For example, when one of the modules stopped reporting and apparently needed to be replaced, Run on Sun immediately contacted Williams to schedule a visit. The rest of the modules were still in full working order, and upon close inspection it was revealed that a connection had come loose. Still, to ensure maximum performance, the company replaced the microinverter at no cost to the school.

Another time Williams received a notification email from Run on Sun after the campus Internet connection had been temporarily cut during some service upgrades. And when the energy dipped from its norm of exceeding system predictions to 98 percent of anticipated, a call came in with a recommendation to check the array for accumulated dirt. After a brief spray with a hose, the system was back to producing at maximum capacity.

At the one-year mark, the school became eligible to receive its first annual rebate from Pasadena Water and Power. This is the first of five annual rebates it will receive, and the dollar amounts are directly correlated to the system's actual production.

When a technician came from the city to assess the energy output of the system, the school was eager to learn the results. The city's readings gave some very welcome news, indeed—the energy generated by the installation was above and beyond the original estimate provided to PWP, and as a result, the first rebate would be larger than anticipated.

"He said, 'You're over your estimate,' and that's all we could ask for," says a thoroughly pleased Williams. "To date, everything that was promised to us was delivered—plus."

In terms of the amount of energy generated, the rooftop system has continued to outpace expectations. The school expected to see a return on its investment in seven years, but it's shaping up to come in as few as six. Because of the installation, Westridge is using 30 percent fewer kilowatt hours and is seeing its bills reduced by thousands of dollars each month, in addition to the rebate. The overall savings is far greater than the cost of running the air-conditioner in the gym, the initial impetus for bringing solar to campus. To Williams, making the decision to go solar was a "no-brainer."

"The neat thing about this is it runs itself. If somebody walks onto campus, they don't know we have a 52 kilowatt solar system on campus," he adds. "They don't see it. It doesn't impact anything. All you do is save money."

Advice to the Solar Reluctant

A true and long-tenured operations man, Williams understands not every business has the same confidence in the value of large-scale solar installations and their ability to save real dollars. Sometimes, people on the ground are convinced of the legitimacy of solar but may run up against superiors still reluctant to make a long-term investment. To those people, Williams gives this advice:

"Don't think short term, think big picture," he says. "Look at the numbers in a way that you're truly aware of the value of sustainable energy and how important it is."

The facilities director advises others in his position to help decision-makers understand that having a solar installation is a way to add value to a company and separate a business from competitors in a way that is meaningful to customers. People will likely support that company by continuing to give it business, especially if the cost savings can be passed on to them.

Addressing the upfront costs is one of the biggest challenges, Williams admits, but it's large-scale projects that offer the best and fastest Return on Investment. When talking to a supervisor, or whoever is responsible for making the decision to invest in a system, speak to the facts. Have your data prepared and on-hand, and be prepared to show concrete evidence that solar would be a worthwhile investment, he says.

Even after a business decides to go solar, there are several more opportunities to remind higher-ups that the decision was a wise one. After installation, data collected from monitoring the system can show the direct benefits to others in the company and create a culture of awareness about the overall benefits of moving toward sustainability. When Williams receives the PWP rebate, he plans to take the paperwork straight to the school's asset management department.

"Once you estimate the legitimacy of it, don't let people forget about it. Remind them it's still working," he suggests.

'Surgere Tentamus'

When Westridge School for Girls was first founded in 1913, women were not even allowed to vote. Despite that fact, founder Mary Lowther Ranney, a well-known architect and teacher way ahead of her time, envisioned a place where young women could rise to new heights. Today, a full 100 years later, Westridge's motto, "surgere tentamus," Latin for "We strive to rise," says a lot about the legacy of its founder.

The decision to make sustainable improvements to the campus stems from that vision, and in that way, Williams says, the new solar installation is the perfect bridge between the school's rich founding principles and its desire to bring state-of-the-art green technology to the campus and the classroom.

He acknowledges how fortunate Westridge was to have students and parents who supported bringing solar to campus and were willing to give funds to the school for completion of the project. Now, when students lead group tours, they point to the south-facing roof of the Fran Norris Scoble Performing Arts Center and proudly let visitors know a fully-operational mini-power plant is running, unseen, above their heads.

Williams encourages others in positions like his to research the many potential benefits of solar. Seriously consider the different providers available, and choose the one who can give you the best value for your investment, because quality means more efficiency, more energy and more savings.

In his case, he acknowledges his own good fortune in finding support for solar. "There are a lot of smart people here," he says. "When you speak intelligently to smart people, good decisions tend to be the result."

VIII. FINAL THOUGHTS

As we have seen, commercial solar power systems can be cost-effective solutions that enhance the reputation of a company while saving it substantial amounts of money at the same time. But these are significant projects that require time and attention if they are to be successfully managed. While there are many high-quality solar contractors out there, sadly, the solar industry is not immune to human foibles of greed or incompetence.

Facility managers and building owners will do themselves a great service if they can follow the advice laid out in this book: take your time to find a series of competent, qualified, "NEARLY perfect" contractors and make sure that you get from each a proposal that makes sense to you. When you have questions, ask. The right contractor—and in the end finding the right contractor is your biggest hurdle—will be more than willing to take the time, and will have the ability, to answer every question that you have.

Find a financing scheme that works for you and your company, whether that is a cash purchase, or something more exotic like Crowd Funding. It is important to know that you have a variety of options, particularly as the size of your project gets larger. Look around, do your homework, and you will find a solution that meets your needs.

Once the project is underway, you will want to remember what we have said here about the steps to expect, and you should plan to be engaged all along the way. Failure to do so can result in a project that perhaps looks good from a distance, but which could be concealing a disaster waiting to happen.

Apply proper oversight, however, and you will have the satisfaction of knowing that you have delivered to your commercial operation a long-term benefit that will not only enrich the company's bottom line, but will improve the world at the same time.

Talk about win-win.

ACKNOWLEDGMENTS

I am indebted to many in producing this book, most especially to my clients over the years and to all who have read my blog (especially those who have commented!). But there are a few without whom this never could have happened, especially:

- My partners, Brad Banta and Velvet Dallesandro, with whom I acquired the experience that is the foundation of this book;

- Andrea Bricco, who contributed the cover photograph and has been such a great supporter of our documentary efforts;

- Sara Cardine, who contributed the Westridge Case Study, interviewing Brian Williams and creating a draft that merged seamlessly with the rest of the book;

- My sister, and biggest fan, Stephanie Jenal, who provided a sounding board, a keen editorial eye over multiple iterations of this manuscript, and constant encouragement; and

- My family, Leslie and Julia, who accepted my decision to leave the well monetized misery of the Law for a risky adventure at happiness and purpose. This book is yet another step on that path; one I never could have trod without them by my side.

Needless to say, any shortcomings that remain are solely my own invention.

GLOSSARY

Term	Definition
3-Phase	A type of *AC Power* delivered by the local utility over three "hot" wires, each one phase shifted by 120°. Commonly used in all but the smallest commercial sites. *Cf., Single-Phase.*
AB 920	California legislation that requires utilities (other than *LADWP*) to make a cash payment to solar customers for any excess energy produced by the end of the annual true-up period. See also, *Net Metering.*
AC Power	Alternating Current refers to power that changes polarity over time. In the United States the standard is to reverse polarity sixty times per second (60 Hz). *Cf., DC Power.*
Accelerated Depreciation	A tax incentive that allows the taxpayer to deduct the cost of an installed solar power system over a shortened period of time. Both California (as well as some other states) and the federal government offer some form of Accelerated Depreciation.
AHJ	Authority Having Jurisdiction, the governmental entity responsible for approving various aspects of a solar project. Frequently includes both the local building and safety department as well as the local utility.

Term	Definition
Azimuth	The orientation of a *Solar Array* relative to true North.
Basic Wind Speed	A factor in calculating *Wind Loads,* the Basic Wind Speed is defined as the fastest three-second gust measured 10 meters above the ground that is not expected to be exceeded but once in fifty years.
BPE	Base Price for Energy, is the underlying price for energy in a *Feed-in Tariff* formula before applying any *ToD Multipliers*.
Capital Lease	A lease that transfers substantially all the benefits and risks of ownership to the lessee. *Cf., Operating Lease.*
CD	Certificate of Deposit.
Crowd Funding	A financing scheme by which a website is used to pair potential investors with desirable investment opportunities. Solar Mosaic is an example of a Crowd Funding website in the solar investment market.
CSI	California Solar Initiative, the sponsor of solar rebate programs for California's *IOUs*.
Data Logger	An electronic recording device that stores data about one or more physical or electrical characteristics of a system, such as peak power, current or temperature.

Term	Definition
DC Power	Direct Current refers to power that never changes polarity, such as that provided by a battery or a *Solar Module*. *Cf., AC Power*.
Demand	A component of some utility *Rate Schedules,* Demand charges are based on the peak power needed by the customer during the billing cycle or other specified period.
Dispatchable	An electrical generation asset that can be turned on or off by the grid operator on command.
EMT	Electrical Metallic Tubing, is thin-wall steel conduit used as a raceway for electrical conductors. *Cf., Rigid Metal Conduit*.
Energy	*Power* consumed over time. Standard units are the Watt-hour (Wh), kilowatt-hour (kWh), Megawatt-hour (MWh), or Gigawatt-hour (GWh). One kilowatt of power (say from lighting ten, 100-Watt light bulbs) used for an hour equals one kilowatt-hour of energy.

Term	**Definition**
Energy Yield	The amount of available solar energy converted into electrical energy by a solar power system. Energy Yield is affected by many factors including: *Solar Module* used, *Inverter*(s) used, *Azimuth, Pitch, Shading, Soiling* and whether the array is a *Fixed-Plate* or *Tracking Array.*
EPBB	Expected Performance Based Buydown, a *Rebate* based on the anticipated (as opposed to actual) output of a solar power system, paid in one lump-sum upon system commissioning. *Cf., PBI.*
Feed-in Tariff	An agreement between a utility and a solar customer whereby the utility agrees to purchase all of the energy produced by the customer's solar power system. None of the system's energy is used to offset the customer's local loads. Cf., *Net Metering.*
Fixed-Plate Array	A solar power system consisting of *Solar Modules* mounted in such a way as to share a common, unchanging *Azimuth* and *Pitch. Cf., Tracking Array.*
FTC	Federal Trade Commission, the federal government agency charged with enforcing, *inter alia,* laws regarding false or deceptive advertising.

Term	Definition
GWP	Glendale Water & Power, the *Municipal Utility* owned by the city of Glendale, CA.
HVAC	Heating, Ventilation & Air Conditioning equipment.
IBEW	International Brotherhood of Electrical Workers, the major electricians' union in the United States that sponsors apprenticeship programs for the electrical trade.
Inverter	An electronic device that, *inter alia*, converts *DC Power* from the *Solar Module*(s) into *AC Power* that can be used to power local loads or feed back to the grid.
Inverter, Central	A Central Inverter is an *Inverter* designed to handle large amounts of power, as much as 1 MW or more. Central Inverters are commonly used in the largest commercial solar power systems.
Inverter, Micro	A Microinverter is an *Inverter* designed to handle small amounts of power and is on the opposite end of the capacity continuum from a *Central Inverter,* with one Microinverter paired with each *Solar Module*. Microinverters are commonly used in residential and smaller to mid-sized commercial solar power systems.

Term	Definition
Inverter, String	A String Inverter is an *Inverter* designed to handle multiple strings of *Solar Modules* wired together in series. String Inverters may have dozens of Solar Modules powering the unit, and, as such, they fall in between on the capacity continuum bounded by *Microinverters* and *Central Inverters*.
IOU	Investor Owned Utility, a utility that is owned by shareholders. *Cf.*, *Municipal Utility*.
IRR	Internal Rate of Return, a common measure of the value of an investment. Technically, it is the interest rate for which the *Present Value* of all of the future cash flows associated with a given project is equal to the cost of the project. The higher the value, the more desirable the investment.
ITC	Investment Tax Credit, a tax credit on an entity's federal income tax, worth 30% of the cost of a solar power system until 2017 when it is scheduled to decline to 10%.
Kilowatt (kW)	One thousand Watts, the most common measure of electrical power.

Term	Definition
Kilowatt-hour (kWh)	One *Kilowatt* of power consumed over one hour of time, it is the most common measure of electrical energy.
LADWP	Los Angeles Department of Water & Power, the *Municipal Utility* owned by the city of Los Angeles, CA.
LCoE	Levelized Cost of Energy, the cost of energy from a solar power system over the lifetime of that system, measured in $/kWh. When the LCoE equals the cost of energy from the local utility, the solar power system is said to be at "grid parity."
LEED	Leadership in Energy & Environmental Design, a building certification program sponsored by the U.S. Green Building Council.
Lien	The legal right of a creditor to sell the collateral property of a debtor who fails to meet the obligations of a loan contract. As part of *PACE* financing, the city or county issuing the bonds will place a lien on the property parcel associate with the solar power system.

Term	Definition
Line-Side Tap	A means of connecting a solar power system to a utility-provided electrical service on the "line" (as opposed to the "load") side of the main electrical disconnect. A Line-Side Tap is required if the current being delivered by the solar power system exceeds the 120% of busbar-rating rule in the *NEC*, or if there simply is no place to install an appropriate circuit breaker on the load side.
Municipal Utility (Muni)	A utility owned by a city, county or other governmental entity. *Cf., IOU.*
NABCEP	North American Board of Certified Energy Practitioners, the preeminent certification entity in the solar industry.
Nameplate Rating	See *STC Rating.*
NEC	National Electrical Code, guidelines governing all aspects of electrical work, including solar power systems, that can be adopted by a local AHJ, and once adopted, become the basis for permitting and inspection.

Term	Definition
Net Metering	A billing arrangement between a solar customer and their local utility whereby energy in excess of that needed by local loads is exported to the grid for credit that is "netted" against energy drawn from the grid.
NREL	The National Renewable Energy Laboratory in Golden, Colorado.
O&M	Operations and Maintenance, refers to the expenses associated with keeping a solar power system running over time and covers everything from cleaning the *Solar Array*(s) to *Inverter* replacement.
Operating Lease	A lease by which the lessee acquires the right to use the leased equipment for a limited time in exchange for periodic rental payments. *Cf., Capital Lease.*
PACE	Property Assessed Clean Energy, a means of paying for a solar power system through the site owner's property taxes.
PBI	Performance Based Incentive, a *Rebate* based on the actual output of a solar power system, paid over several (usually five) years. *Cf., EPBB.*
PE	Professional Engineer.

Term	**Definition**
Performance Meter	A meter of varying degrees of accuracy that measures the output of a solar power system. Performance meters can be as simple as a refurbished analog meter or as sophisticated as a *Revenue-Grade Meter*.
Pitch	The inclination of a *Solar Array* in degrees relative to horizontal.
Power	The ability to do work. Standard units are the Watt (W), kilowatt (kW), Megawatt (MW), or Gigawatt (GW).
PPA	Power Purchase Agreement, a long-term contract to purchase energy from a solar power system owner at a defined cost which may rise over time. *Cf., Operating Lease.*
Present Value	The value, as of a specified date, of future economic benefits and/or proceeds from sale, calculated using an appropriate discount rate.
PTC Rating	The measure of a *Solar Module's* power output based on Photovoltaic Test Conditions. It is always lower than the module's nameplate or *STC Rating*. *Cf., STC Rating*.
PTO	Permission to Operate, an official notice from the local utility that a solar power system has been approved to begin regular operations connected to the grid.

Term	Definition
PVWatts	A computer model created and maintained by *NREL* for analyzing the performance of solar power systems. PVWatts is the model commonly used for creating system performance projections used by utilities in determining solar *Rebates*.
PWP	Pasadena Water & Power, the *Municipal Utility* owned by the city of Pasadena, CA.
Rebate	A payment from a utility to a solar customer as an incentive to install a solar power system. See also *EPBB* and *PBI*.
Revenue-Grade Meter	A *Performance Meter* designed to measure the output of a solar power system to a level of accuracy deemed sufficient by the local utility.
Rigid Metal Conduit	A thick-wall steel conduit used as a raceway for electrical conductors and providing greater conductor protection than *EMT*. *Cf. EMT*.
ROI	Return on Investment, a calculation that determines how soon a particular investment will be paid back (the "payback period") based on a series of anticipated cash flows.
SCE	Southern California Edison, a large *IOU* in the greater Los Angeles region.
SE	Structural Engineer.

Term	Definition
Shading	The amount of sunlight blocked from reaching the solar array due to obstructions such as trees, buildings, power poles, etc. Ideal sites have no shading; acceptable sites will be 90% shading free. Sites with shading values below 90% should not be built without the use of *Microinverters* or other comparable technology. See also, *Shading Analysis*.
Shading Analysis	An analysis that determines the degree of *Shading* at a site. Typical output of such an analysis is a month-by-month assessment of how much of the available sunlight will actually reach the array. (An unshaded site would be at 100% for each month.)
Single-Phase	A type of *AC Power* supplied by the local utility where power is delivered over two "hot" wires, in phase with each other. Commonly used in residential electrical systems or very small commercial sites. *Cf., 3-Phase*.
Site Plan	A series of drawings that shows where a solar power system is located at a given site and may include detail drawings showing how various system components are mounted or connected together.

Term	Definition
SLD	Single-Line Drawing, a simplified schematic view of the electrical components involved in a solar power system. Typically also includes specification of conductors and conduits to be used in the system.
Soiling	Term for the dirt, debris and other deposits that accumulate upon *Solar Modules* over time, thereby reducing their *Energy Yield*.
Solar Array	A collection of *Solar Modules* with a common *Pitch* and *Azimuth*.
Solar Cell	A semiconductor device that converts light (photons) into electricity. Multiple Solar Cells are combined together to produce a *Solar Module*.
Solar Module	Also referred to as a solar panel, an integrated electronic device consisting of multiple *Solar Cells*, arranged in an interconnected grid, encapsulated against moisture and other environmental agents, and held together by a rigid frame for mounting.
Solar Noon	The moment in the day when the Sun is at its highest point in the sky.
Solar Pathfinder	A device used to perform a solar *Shading Analysis*.

Term	Definition
STC Rating	Also known as the *Nameplate Rating*, the measure of a *Solar Module's* power output based on Standard Test Conditions. It is always higher than the module's *PTC Rating*. *Cf., PTC Rating.*
Tariff	See *Utility Rate Schedule*.
ToD Multipliers	Time-of-Delivery Multipliers; in a *Feed-in-Tariff* formula ToD Multipliers are applied to the *BPE* to determine the price paid for a quantity of energy at a particular day and time.
TOU	Time-of-Use, a term used to describe *Utility Rate Schedules* that charge different amounts for energy consumption and/or peak power demand, based on the time of day, or a period (such as a season) during the year.

Term	Definition
Tracking Array	A solar power system consisting of *Solar Modules* mounted on a frame that can change its *Azimuth* over the course of a day to track the Sun, and its *Pitch* over the course of the year to account for variation in the Sun's path. A *Tracking Array* that only changes *Azimuth* is called a single-axis tracker, whereas one that also changes *Pitch* is called a two-axis tracker. Although *Tracking Arrays* can increase annual *Energy Yield* by as much as 45% over a comparable *Fixed-Plate Array*, they are not commonly used in rooftop commercial solar power systems due to space and maintenance constraints.
Tranche	A division or portion of a pool (as in an investment pool).
Transformer	An electrical device used with *AC Power* systems that can increase ("step-up") or decrease ("step-down") the voltage in a circuit. Transformers are commonly used with an *Inverter* to match its output voltage to that of the electric service provided by the local utility.
Usage	The most common component of electric utility bills, based on the total amount of grid-provided energy consumed during a billing cycle.

Term	Definition
Utility Rate Schedules	The formula by which a utility calculates a customer's bill. Utilities, especially *IOUs*, have many, many rate schedules, some of which will be more or less advantageous to solar power system owners.
Wind Loads	Wind-driven force that attempts to lift a solar array from its mounting. For a roof-mounted system, the uplift of wind loads is resisted by a variety of means including ballast (on a flat roof) or bolted or screwed in attachment plates (on pitched roofs).

INDEX

ABOUT THE AUTHOR

Jim Jenal founded Run on Sun, a solar power installation and consulting company in Pasadena, California, in 2006. As its CEO, he has authored the company's blog, *Thoughts on Solar*, since 2009, and he has been quoted on solar policy and the state of the solar industry in media such as the *Los Angeles Times*, the *Wall Street Journal*, and the *Guardian*, as well as in numerous trade publications.

A NABCEP Certified Solar Installation Professional, Jim also holds degrees in Mathematics, Computer Science and Law.

Jim lives in Pasadena with his wife, Leslie, their daughter, Julia, and the exceptionally scruffy dog, Marshmallow.

23273074R00093

Made in the USA
Middletown, DE
20 August 2015